David Soal £40.00

C287.

Aero.

£20:

Sale

Flight simulation

CAMBRIDGE AEROSPACE SERIES

Flight simulation

Edited by

J. M. ROLFE
Senior Principal Psychologist, Ministry of Defence

K. J. STAPLES
Royal Aircraft Establishment, Bedford

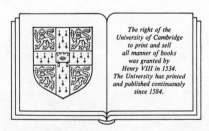

The right of the
University of Cambridge
to print and sell
all manner of books
was granted by
Henry VIII in 1534.
The University has printed
and published continuously
since 1584.

CAMBRIDGE UNIVERSITY PRESS

Cambridge

London   New York   New Rochelle

Melbourne   Sydney

Published by the Press Syndicate of the University of Cambridge
The Pitt Building, Trumpington Street, Cambridge CB2 1RP
32 East 57th Street, New York, NY 10022, USA
10 Stamford Road, Oakleigh, Melbourne 3166, Australia

First published 1986

Printed in Great Britain at The Bath Press, Avon

British Library cataloguing in publication data
Flight simulation.
1. Flight simulators
I. Rolfe, J. M.   II. Staples, K. J.
629.132'52'078   TL712.5

Library of Congress cataloguing in publication data
Flight simulation.
Bibilography
Includes index.
1. Flight simulators, I. Rolfe, J. M.
II. Staples, K. J.
TL712.5.F54   1986   629.132'52   85-28083

ISBN 0 521 30649 3

TM

Contents

To the memories of Wladyslaw Sluckin (late Professor of Psychology, the University of Leicester, England) and Svante Skans (late Director, LUTAB, Sweden). Two wise friends whose enthusiasm and encouragement is embodied in this book.

Preface

The idea of producing a textbook to introduce the subject of flight simulation is a long held ambition we have both cherished. Three influences led to its becoming a reality; the encouragement of our fellow enthusiasts in the Flight Simulation Group of the Royal Aeronautical Society, the initiative of the Cambridge University Press in commissioning the book to form one of the first titles in its revived Aerospace Series and the willingness of the Ministry of Defence to allow us to prepare the book, while wishing it to be understood that they do not necessarily endorse its contents.

We are well aware of the wide range of scientific and engineering disciplines encompassed by flight simulation. We have sought to provide a sound basic description of each of the many facets of the subject, with bibliographical pointers to further, more detailed information. However, in view of the diversity of topics, we also thought it prudent to enlist the help of specialist contributors to ensure factual accuracy. We are indebted to the following friends and colleagues for their contributions to the chapters of this book.

Arthur Barnes (Chief Simulation Engineer, British Aerospace, Warton Division, England) for the chapter on the use of flight simulators in research.

Martin Bolton (Lecturer in Microelectronics at Bristol University, England) for the chapter on the history of flight simulation.

John C. Dusterberry (formerly of the NASA Ames Research Center, California, USA) for the chapter on motion systems.

Brian P. Hampson & Maxwell L. Rutherford (respectively Manager, Project Engineering and Manager Flight Simulator Systems Engineering, CAE, Montreal, Canada) for the chapter on the integration, testing and acceptance of simulators.

Brian Lynch (Chief Engineer, Flight Simulation Division, Singer Link-Miles, Lancing, England) for the chapter on simulator structures and cockpit systems.

John H. Marsden (Manager, Systems Engineering Department, Rediffusion Simulation, Crawley, England) and his colleagues for the chapter on the simulation of aircraft systems.

Malcolm E. Roberts (Rediffusion Simulation, Crawley, England) for the chapter on the design of instructor facilities.

A. Michael Spooner (formerly Head of Advanced Simulation Concepts Laboratory, Naval Training Equipment Center, Florida, USA) for the chapter on visual systems.

Barry Tomlinson (Head of Advanced Informatics Research and formerly Head of Flight Simulator Trials, Royal Aircraft Establishment, Bedford, England) for the chapter on mathematical models for flight simulation.

Bill Wooden (formerly Chief Inspector of Flight Operations, UK Civil Aviation Authority) for the chapter on the use of flight simulators in training.

The result of this international partnership is contained in the following pages. It encapsulates the spirit of collaboration and good fellowship that abounds within the flight simulation community.

John M. Rolfe & Ken J. Staples
May 1986

1

An introduction to flight simulation

1.1   Introduction

While not neglecting the applications of flight simulation, the principal objective of this book is to provide an introduction to the fundamentals of flight simulation and the elements that contribute to the construction and operation of the modern simulator.

As its name implies the object of flight simulation is to reproduce on the ground the behaviour of an aircraft in flight. The practical value of flight simulation can be judged by the extensive use of the technique in aerospace research and development and by the fact that more than 500 flight simulators are in use, throughout the world, for training and maintaining the skills of civilian and military aircrew.

When used for research, flight simulators allow designers to explore the implications of different design options without having to incur the expense and delay arising from building and testing a range of prototypes. Examples are the extensive use of flight simulation in the design and testing of the Concorde supersonic airliner and the role played by simulation in the manned space programme.

Flight simulation has provided a means of evaluating the likely behaviour and consequences arising from abnormal operating configurations. Solutions to handling problems associated with deep stall, clear air turbulence and, recently, wind shear have all been worked through with the aid of simulators.

Flight simulation has established for itself a significant position in the ab initio and advanced training of aircrew. The market for simulators is world wide and annual expenditure on new devices involves millions of pounds. Nevertheless, hard-headed commercial operators and efficiency conscious air forces are prepared to argue five major advantages accruing from the use of simulators in training. These are:

increased efficiency, as training will not be interfered with by factors such as adverse weather conditions, airspace limitations or aircraft availability

increased safety whilst training, coupled with the ability to control the level of task demand applied during training

lower overall training costs

the facility to practise situations which for reasons of expense, safety and practicability cannot be rehearsed in the real world

the reduction in operational and environmental disturbance.

These successful applications of flight simulation can be used as the justification for the expenditure of thought and other precious resources on arriving at a greater understanding of the factors which go to make for effective simulation. Whilst accepting this to be so it may also be argued, with justification, that flight simulation, as an area of knowledge and technology, is independent of its potential applications. Scans & Barnes (1979) pointed to the fact that, while flight simulators have existed for more than half a century, most of the advantages and benefits from simulation have only occurred since the mid 1960s.

The aim of setting out to create a representation of flight which is indistinguishable from the real thing has been a sufficient challenge and for some practitioners in simulation the pursuit of this goal has earned them the facetious title 'the kings of toyland'. Nevertheless flight simulation is worthy of study for its own sake for, as later sections of this book will show, it has resulted in greater understanding of the process of manned flight and acted as the stimulus to innovation and invention in fields as disparate as computer graphics, hydraulics and human skills analysis.

1.2   Definitions

Simulation can take a variety of forms and it is therefore essential to define the scope of this book. A common feature of all simulations is that they attempt to provide an operating imitation of a real activity. Just how this is achieved will vary with the nature of the simulation. The economist and operational analyst may create numerical simulations totally within a computer while the earth scientist or hydrographer may use dynamic physical representations of parts of the environment. These two examples represent the different levels of abstraction which can be contained within the use of simulation. A further dimension is the level of human involvement that occurs within the simulation. One extreme is represented by the computer simulation case while the other by management, or command and control decision making simulations which involve people as participants in and controllers of, the simulation.

This book is about the design and operation of devices with low levels of abstraction and high levels of human involvement whose purpose is to simulate the behaviour of an aerospace vehicle. Aviation does use other forms of simulation but these will not be dealt with in any detail here. The essential form of flight simulation is the creation of a dynamic representation of the behaviour of an aircraft in a manner which allows the human operator to interact with the simulation as a part of the simulation.

The form of simulation dealt with here involves the combination of science, technology and art to create artificial realism for the purpose of research, training and pleasure. This statement identifies two streams of influence in the development of flight simulation. Firstly, the creation of simulations is as much an art as it is a science and technology. Secondly, the purpose for which the simulation may be employed can be both creative i.e. improve aeronautical design and operating proficiency, and recreational, i.e. provide enjoyment by having created a simulation which can give pleasure and satisfaction.

The streams of influence and application referred to above can be seen to be present in the historical development and current state of flight simulation. As an example of the combination of art, science, and technology to achieve effective simulation Tabs (1964) described how, at the start of the Second World War, theatrical set designers were called upon to collaborate in designing training simulators for navy torpedo attack pilots operating against enemy shipping.

Still in the context of the evolution of flight simulators an example of the pleasure or purpose dichotomy is the dilemma faced by Edwin Link, the inventor of the Link Trainer, as to whether his device would have more attraction to flying schools or amusement parks (Kelly & Parke 1970).

These design and use features remain as influences in the present day context of flight simulation practices and applications. In the development of new techniques, particularly those associated with the representation of the visual world using computer generated graphics, the influence of the artist in representing texture and accurate patterns of light and shadow is clearly important. At the same time the advent of the micro computer and its ready adoption as an adult toy has resulted in the creation of flight simulator game packages that offer recreation and challenge but do not set out to teach the user to fly.

Clearly then, flight simulation presents a multidisciplinary challenge to skill and ingenuity. This position has been described most succinctly by a former Chief Test Pilot of the British Civil Aviation Authority (Davis 1975).

> Flight simulation is a fascinating and challenging field of science. It encompasses a wide field of disciplines from hydraulics and

electronics to aerodynamics, performance and optics as well as a massive contribution in terms of human engineering. The biggest challenge is the overcoming of the limitations of not being able to establish all the root equations of motions as fundamental parameters and the need to cook the system so that these compromises are not significant. In many ways working with simulators is much more demanding than working with real aeroplanes.

1.3   The basic structure of flight simulation

A simulation may be considered as comprising of three parts. These are, a model of the system to be simulated, a device through which the model is implemented and an applications régime in which the first two elements are combined with a technique of usage to satisfy a particular objective.

A model as its name implies is a representation, actual or theoretical, of the structure or dynamics of a thing or process (Abt 1964). Models may take one of three forms: physical, analog or linguistic. An example of a physical model is a scale model of an aircraft used for aerodynamic tests in a wind tunnel. An example of an analog model is a recorded sequence of electrical signals representing turbulence. This is used to vary the amplitude of movement of a bank of hydraulic jacks moving a specimen wing section of an aircraft during fatigue tests. The most usual form of linguistic model encountered in simulation is the mathematical description of the behaviour of a system in terms of a number of equations. A mathematical representation of an aircraft and its dynamic response forms the basic model used with contemporary flight simulators. In every one of the three model forms identified above a wide variety of models will be found. For example, verbal models of aircraft operations can play a useful role in teaching a student procedural and emergency drills. These linguistic problem-setting models put the student into a decision-making situation in which he must decide what course of action to take in response, for example, to a fire warning light in the cockpit. While the above classification of models has a wide acceptance there is also agreement that the boundaries between types are diffuse and that they form a continuous spectrum starting with exact physical mock ups and extending to completely abstract mathematical models.

The operating device is the element in the simulation through which the model is operated and evaluated. As the above examples show the operating device can take any number of forms ranging from a wind tunnel to a programmed book. However, in the context of this book the operating device will be the flight simulator and it will, inevitably, appear as the most apparent facet of the simulation. However, the model is essential. Without an adequate or representative model the simulator would either be inert or inaccurate.

The operating régime is the application for which the simulation is intended. Once again there are a variety of applications. Elmaghraby (1968) has listed five common uses of models and simulations:

(1)   as an aid to thought
(2)   as an aid to communication
(3)   for purposes of training and instruction
(4)   as a tool for prediction
(5)   as an aid to experimentation

The need to pay attention to the technique of usage is important because the simulation will rarely be a perfect representation of reality. The task, or set of functions, for which the simulation is used is thus constrained and elements of the real-life task may themselves have to be distorted to allow a valid solution to the aim of the simulation.

The interrelatedness of the three elements representing a simulation is not sequential, with the model determining the choice of facility and the facility deciding the application. The elements will be looked at in this order in the following chapters of this book because it forms a logical way of looking at the elements of flight simulation. The ability to create an effective model, and the nature of the model, will have a direct influence upon the choice of facility to operate the model. But, it is also important to stress that when a simulation is being created for a specific application the intended use must be taken in account when selecting the model upon which to base the simulation.

Decisions about the choice and content of the model and the associated device are an aspect of that facet of simulation practice which has already been described as being in some part an art. Shannon (1975) emphasises this as a key point in simulation when he says:

> The tendency is nearly always to simulate too much detail rather than too little. Thus, one should always design the model around the questions to be answered rather than imitate the real system exactly. Pareto's law says that in every group or collection there exists a vital few and a trivial many. Nothing really significant happens unless it happens to the vital few. The tendency among systems analysts has too often been to transfer all the detailed difficulties in the real situation into the model, hoping that the computer will solve their problems. This approach is unsatisfac-

tory not only because of the increased difficulty of programming the model and the additional cost of longer experimental runs, but also because the truly significant aspects and relationships may get lost in all the trivial details. Therefore, the model must include only those aspects of the system relevant to the study objectives.

Clearly Shannon's advice is directed toward those who would use simulations for research. However, in relation to the use of flight simulators for training his view is complemented by Prophet, Caro & Hall (1971) who remark that:

If the aircraft is a poor learning environment, and for many flight related skills it is one of the poorest imaginable – then a ground based duplicate of that environment will not necessarily be a better one in which to learn.

The above advice makes plain the point that not only must a thorough examination of the application be undertaken before deciding upon simulation as the solution but also that the choice of what elements to include in the simulation must be made with care. If these steps are not taken then it is possible to expend much time and effort on creating simulations which do not meet the requirements. The following three examples show how simulations can fail to meet expectations; they should not be thought of as bad simulations, rather as demonstrations of how hard it is to produce a simulation which meets requirements.

1.3.1   *The effect of an inadequate model*

At one stage in the development of the Concorde supersonic airliner, interest was centred upon the aircraft's operational integration into air traffic control patterns. One task was to find out how Concorde could be fitted into traffic flow if, because of an engine failure, it was necessary to descend to a lower altitude and into the airspace normally occupied by subsonic aircraft. To examine this a complex exercise was set up in which a trans-Atlantic flight was flown in the Concorde simulator. This flight was to be injected into actual trans-Atlantic air traffic patterns. In essence a ghost Concorde was to fly across the Atlantic. To achieve this a link was established between the Concorde simulator and the air traffic control computers handling the actual movements of aircraft between Europe and the United States.

The scenario for the exercise was that an experienced flight development crew from one of the airlines which would be operating Concorde would fly the simulator. They were not told that an engine failure would be simulated soon after they had reached cruising altitude to the west of Ireland. At that point the expectation was that the captain would declare an emergency and ask to be allowed to descend in order to continue the flight at subsonic speed, something the aircraft was capable of doing. This was the point at which lessons were to be learned and the object of the simulation.

The exercise commenced and at the correct point in the flight the simulator operator injected the engine failure. The captain of the Concorde reacted immediately to the incident by taking all the necessary steps to close down the engine and prevent a fire. He then declared his situation to ATC and asked for assistance to return to Heathrow Airport at London. This he did.

In the subsequent debrief the captain was asked why he did not continue with the trans-Atlantic flight at subsonic speed and lower altitude. His answer was that, while this was possible, he knew there were no spare engines at the diversion airfield in North America. He therefore decided that the correct course of action was to return to London so that a new engine could be fitted. In this way the aircraft would be back in service with less delay.

The incident did not invalidate the objects of the study and it produced an outcome which, although not what was expected, was nevertheless a valid one. It emphasised the importance of ensuring that the model for the simulation contained all the relevent information required to produce an appropriate simulation. When the added factor of engine availability was understood it was possible to repeat the exercise with the crews knowing in advance that simulated spare parts were present on the other side of the Atlantic.

1.3.2 *Effect of inadequate facilities*

The 1960s saw the introduction of the commercial passenger carrying jets such as the Boeing 707. These aircraft flew higher than their predecessors and their operators for the first time encountered the effect known as clear air turbulence. This turbulence was not associated with any observable cloud condition, it was violent and it could generate severe handling problems. Crews were trained to deal with the turbulence in flight simulators which, at that time, were fixed base devices, that is, they did not provide any physical motion cues. In one incident an American airliner encountered clear air turbulence and descended 'quickly' through 25000 feet before the crew were able to regain control. The aircraft was fitted with a flight data recorder and it was possible to reconstruct the pattern of disturbance leading up to the loss of control. The flight was then reflown in a training simulator using a number of experienced flight crews. They had no difficulty in dealing with the situation without losing control of the aircraft. The flight was then reproduced on a sophisticated research simulator which could simulate the physical motions associated with the turbulence. Now a marked change in the crews' performance took place. The motion created a false illusion of climbing when the aircraft was in fact already descending. As a result the crews moved the controls in the wrong direction and increased the rate of descent. The study showed that, when motion was present, the crews in the simulator behaved in the same way as the crew in the aircraft. However, this was only so for the first run in the simulator. Once the disturbing motion effect has been experienced and understood the crews adjusted their behaviour to deal appropriately with the situation. The lesson learnt from this study was that for some levels of simulator training the presence of disturbance motion cues is essential if effective training is to take place. Subsequent to that study it became widely accepted that training simulators should be fitted with motion systems.

1.3.3   *The effect of inappropriate application*

Other areas of application have their problems as well as flight simulation. This example is taken from the use of simulators to train the masters of supertankers to familiarise themselves with manoeuvering the vessels. For this purpose a major oil company which operated supertankers constructed a 1 : 25 scale representation of an oil terminal and its approaches. In this scale world tanker masters could manoeuvre realistic models of supertankers constructed to the same scale and large enough to accommodate the trainee master. However, in order to give the operator effective control over the scale model it had to move about the 1:25 world with a timescale of 5 : 1 compared with the full size tanker. When a validation study was conducted it was found that training in the 5 : 1 timescale has no positive transfer of skill to the 1 : 1 situation and that incorrect relationships between command and response were being learned.

In this case the lesson is an important one that methods of simulation appropriate for some applications may not be suitable in others. As a result of the study referred to above training was moved from the reduced scale device to a full-scale simulation in a ship-handling simulator.

1.4   The flight simulator: models and facilities

The earliest attempts at flight simulation employed devices which demonstrated the effects of control movement on the behaviour of an aircraft. To achieve this a rudimentary representation of an aircraft was mounted, out of doors, on a system of pivots. It relied upon airflow over control surfaces generated by the wind to create and sustain the simulation. In this case there was a limited pragmatic model with the essence of the simulation being in the device, the tethered airframe and in the environment (Haward 1910).

The next chapter will review the technological history of flight simulation. One important point is that the model has gone through a number of transformations before arriving at the mathematical representation of flight dynamics which is the basis for contemporary simulation techniques. The ability to express the behaviour of an aeroplane and its major internal systems by means of a series of equations existed long before an implementing device was available to rapidly translate mathematical calculations into a representation of flight. This became possible with the advent of the computer. When this capability presented itself two ensuing developments resulted. Firstly, the scope and magnitude of the model could be greatly extended. Secondly, because the model was being implemented using a device which used electrical pulses, initially in analogue and subsequently in digital form, it opened the way to the development of enhancements to the implementing device.

Consider first the extension of the model. The aircraft model is the nucleus of the simulation and within it there will be contained submodels representing the handling characteristics of the airframe, engines and manual and automatic control systems. Also present will be a further range of submodels for display and communications systems. Lastly the aircraft model will contain a subset relating role functions; for example with a fighter aircraft there will be weapons models.

For the simulation, the aircraft model must operate within a simulated environment. The extent of that environment will be determined by further models. Once again the scope for modelling has changed significantly. In early simulations the simulated aircraft operated in a relatively un-structured and limited environment. Flying was done on instruments and the dimensions of the world were represented by radio beacons and other fixed location navigational aids. With the capabilities of the modern digital computer it is possible to create detailed air and ground environments within and over which the aircraft model can be translated.

The air environment model will contain representations of meteorological conditions and other aircraft which may be seen and heard. The ground environment model will contain a terrain model which will be enhanced by specific features: models of towns, arrival and departure airfields and, for military applications, potential targets. A radio navigation model will be expanded to include an inertial navigation model and an air traffic control model.

A simple representation of the above formulation of models for flight simulation is shown in Fig. 1.1.

It will be obvious that while the three component models are identified separately they must work together in order to create a representative simulation. For example the aircraft model will compute height, heading and attitude. This information will be used by the air environment model together with its own weather conditions model to determine the slant range visibility to the ground. In turn this data will be used in

Fig. 1.1. Diagramatic representation of the three principal models contained within a flight simulation.

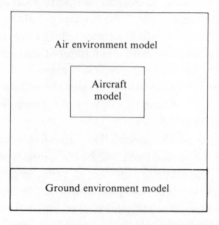

Fig. 1.2. Simple representation of the principal elements of a flight simulator.

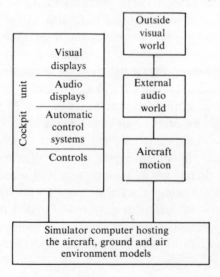

conjunction with the ground model to determine what terrain features can be seen from the aircraft.

The second advantage that accrues from the use of a computer as the host for the simulation model is the ability to extend the sophistication of the implementing device. The extent of this expansion is shown in a simple form in Fig. 1.2.

Early simulation was confined to reproducing the dynamic behaviour of the aircraft via the basic controls and displays within the cockpit. However, with increased computing capacity and improved electromechanical facilities much higher levels of simulation are possible in terms of realism, or fidelity with the real world.

Inside the cockpit the simulator engineer has been able to keep up with developments in aircraft instrumentation and avionics. This has been aided by the greater use of electronics in the cockpit so that differences between the simulated and the actual instrument have been reduced. For example in the era of the direct pressure-driven altimeter the simulator engineer had to construct a totally different drive mechanism for the simulated altimeter. With the advent of the servo-driven and computer generated altitude displays it is possible to use the actual displays driven by electrical signals generated by the interface to the simulator computer instead of the air data computer.

The ability to reproduce what goes on inside the cockpit and how the aircraft handles under manual and automatic flight control is one facet of the increased capability available to the simulator designer. The second, and equally as significant a development, is the ability to reproduce in greater breadth the behaviour of the aircraft in relation to the external world. This capability shows itself in three major areas: the provision of cockpit motion, and the representation of both the external visual and audio environments.

The motion of an aircraft in flight can be considered in relation to the three rotational axes and the three translational axes. Motion systems providing movement along all six axes are now in use. Although opinion is divided as to the extent of the motion that is really essential there is agreement that motion plays an important role in the simulation of flight.

Equally challenging is the ability to provide a representation of the external visual scene from the cockpit. The task is to provide sufficient detail together with both field and depth of view to allow a pilot to perform tasks which require rapid and accurate assimilation of information from the outside world. Examples of situations where an accurate representation of the external world can be vital are for visual approaches and landings and for very low level 'nap of the earth' flying in military helicopters. It is also important to be able to degrade the visual scene in order to simulate poor visibility and other weather conditions.

The provision of motion and visual cues for flight simulators highlights another aspect of the designer's task. This is that, as the devices are intended for use with human operators, a knowledge of the physiological and psychological factors influencing the sensation and perception of motion and visual stimuli is important. Thus another knowledge dimension is introduced into the simulation arena.

The external audio environment is important for its ability to create the illusion of the presence of other traffic competing for both airspace and audio transmission time – a contributor to aircrew workload. The problems with the simulation of this aspect of flight are not as great as those of the motion and external visual dimensions. Nevertheless there are unique features which present their own difficulties. One aspect of the audio environment is the representation of aircraft-generated noises. The most obvious of these is engine noise, but tyre squeal, runway rumble, undercarriage retraction/extension and locking and pre-stall noise are examples of sounds that must be reproduced. These noises not only convey important information to the crew, they also mask noises generated by components of the simulator, for example the motion system. A major problem is the ability to create a flexible and responsive audio transmissions scenario which can adapt to unexpected changes in a simulated flight. One way of achieving this is to have simulator staff play the role of other aircraft and air traffic controllers. This labour intensive remedy is only partially successful and the solution may lie with the application of voice synthesis technologies.

The representation of the basic elements of the flight simulator, as they appear in Fig. 1.2, shows only those features which contribute to the simulation of a particular vehicle. An important additional part of the facility, which is not present in the aircraft being simulated, is the provision of a control station from which the simulation will be operated. The form the controllers' station will take is dependent upon the role of the simulator. If the device is a training simulator the instructor will need to have a workstation that allows him to perform the functions of providing training as well as being able to set up and control the behaviour of the simulator.

The elements that comprise the flight simulator require specialised design skills and fundamental knowledge. Equally important and specialised is the ability to integrate all the subelements to create a device which not only looks like an actual aeroplane from inside the cockpit but which also responds like it. Moreover, if the simulator is to be used for training in the commercial aviation field the process of element integration is one which will not only have to satisfy the demands of the intending purchaser but also those of a legislative authority, such as the Civil Aviation Authority in the United Kingdom, who will have the responsibility for approving the use of the simulator for training.

The major part of this book will be devoted to a consideration of the methods of modelling flight and the means of implementing the model to create a flight simulator. The third aspect of the subject is the application of simulation. While it was stated earlier that this topic was not the principal purpose of the book it does deserve some attention. The potential value of a device which could represent the behaviour of an aircraft in flight on the ground was recognised more than sixty years ago by Reid & Burton (1924). They concluded that such devices, if they could be constructed, could be used to:

(1)   test the ability of subjects to fly and land successfully

(2)   assess the rate of acquisition of flying skills

(3)   train pupils on those particular coordinations necessary for aircraft control

(4)   classify subjects for different forms of flying service

All of these predictions have been fulfilled and flight simulators play an effective part in flying training. The flight simulator's value as a research tool is also recognised. The applications of simulators, in research and training, will be considered in later sections of the book. However, the objective in the present context will not be to provide a comprehensive analysis of applications but, rather, to identify some of the fundamental issues which have to be taken into account when seeking to utilise flight simulations.

The editors hope that this book will convey an appreciation and understanding of the multidisciplinary nature of flight simulation, science, technology and art. Other authors have drawn comparisons between the skill of the stage illusionist and the flight simulator specialist. There are undoubted similiarities, both seek to deceive the senses and to convince the observer that what is being experienced is real rather than a simulacrum of reality. However, the major and fundamental difference is that whereas the illusionist performs to an audience the practitioner in flight simulation must convince human participants who are themselves part of the simulation. Therein lies the challenge, the reward and the satisfaction.

2

A short history of the flight simulator

2.1 Introduction

An account of the development of the technology of flight simulation would not be as narrow in scope as it would at first sight appear. Flight simulation has made immediate use of advances in technology throughout its 75-year history and indeed, its demands have more than once spurred new developments in the supporting technologies.

These demands have always been to produce as faithful a reproduction as possible of the behaviour of an aircraft – both of types already in use and of those still in development. This short review will focus on the evolution of techniques for flight simulation using for illustration examples of devices which best represent the major steps in their evolution. The majority of these were built for training purposes, although of course flight simulation includes machines built for engineering and psychological research. Most of these latter are omitted from this account for reasons of space, as are simulators built for training in the non-piloting tasks of aircrews, such as navigation, radio operation and gunnery.

2.2 Early efforts

Before the first public flight by the Wright brothers in 1908, it was widely believed that a powered aeroplane would be as stable a vehicle as an airship, one which could be flown without previously acquired skills (Hooven 1978). However, their aeroplane depended on the operator for its equilibrium – flight would be impossible until a particular set of skills had been learnt. The successors of the Wright aeroplanes were just as difficult to fly, each one having its own, potentially fatal, idiosyncrasies (Robinson 1973). A logical approach to teaching at least some of the skills required was to use an example of the real aircraft, but safely linked to the ground. An early example of this was the Sanders Teacher, introduced in 1910. A piece in *Flight* stated: (Haward 1910)

> The invention, therefore, of a device which will enable the novice to obtain a clear conception of the workings of the control of an aeroplane, and of the conditions existent in the air, without any risk personally or otherwise, is to be welcomed without a doubt. Several have already been constructed to this end, and the Sanders Teacher is the latest to enter the field.

The Sanders Teacher was a modified aeroplane mounted on a universal joint linked to the ground. In it the student could learn the control movements necessary to maintain equilibrium. Of course such a device depended on a satisfactory supply of wind and relied on gusts to produce disturbances. Another example was that of the Italian Gabardini company, who produced a captive version of their monoplane, based on the same principle, for teaching the use of the controls (Janes 1919). The idea of using a captive aeroplane for elementary training or for amusement was patented in Britain by Eardley Billing (Billing, 1910). His device, which was available at Brooklands Aerodrome, was in fact not an adapted aeroplane but a purpose-built machine (Fig. 2.1). The control column operated planes

Fig. 2.1. A trainer of the Billing type. (Courtesy *Flight International*.)

which enabled equilibrium to be kept and the rudder bar was connected to the base to allow the machine to be rotated to face the wind.

A variation of this principle, one which did not rely on the wind, was also frequently tried in these early days of flying. These devices relied on an instructor to provide the disturbances while the student would attempt to maintain equilibrium by manipulation of controls connected through wires and pulleys to the base. Walters' machine was of this type (Walters 1910). An improvement on this solved the problem of the student having to act in direct opposition to the disturbance forces applied by the instructor, by adding a second universal joint. In the Antoinette 'apprenticeship barrel' illustrated in Fig. 2.2 two instructors are required and the universal joints are provided by barrels. Even though these contrivances did not require a breeze for operation, their training value must have been even less, or perhaps even negative, due to their immediate response to control movements by the student.

With World War I and the development of military aviation the first requirement arose for teaching rapidly the skills of flying to large numbers of people. Simulation had virtually no impact. In Britain the training system devised by Smith-Barry emphasised actual flying from the earliest stage. In France, where the Blériot system was used, the pupil advanced through a planned evolutionary sequence (Winter 1982). His first experience was at the controls of a 'Penguin', a monoplane with sawn-off wings capable of hopping at about 40 mph down the 'frogs meadow' (Winslow 1917). The penguin idea was well known even before the war, but the French seem to be the only ones to have used it in serious training.

The need to reduce the wastage rate in World War I flying taining encouraged the growth of the discipline of aviation psychology. Many of the tests for flying aptitude developed by these psychologists required the

Fig. 2.2. Antoinette trainer.

development of devices intended to measure performance in certain tasks thought to represent essential flying skills. Two popular types of test were those of reaction time and coordination. In 1915, a machine of the former type was proposed – this consisted of a rocking fuselage fitted with controls and an electrical recording apparatus in which the response of the subject to tilting manually produced by the examiner would be recorded (Anderson 1919). Amongst many other examples, only the work of Reid and Burton will be cited (Reid & Burton 1924). This was an electrically controlled apparatus mounted in a dummy cockpit and which could record the time taken for the subject to restore an attitude display to its central indication.

It was mistakenly believed during this period that the vestibular apparatus enabled a person to sense orientation in the air as well as on the ground (Robinson 1973). It was later realised that orientation depends largely on vision. A device built on the earlier theory was the Ruggles Orientator (Ruggles 1917, 1918). This consisted of a seat mounted within a gimbal ring assembly which enabled full rotation of the pupil in all three axes in addition to providing vertical movement. All motions were produced by electric motors in response to movements of the student's and instructor/examiner's sticks and rudder bars. This device was stated to be useful for 'developing and training the functions of the semi-circular canals and incidentally to provide such a machine for training aviators to accustom themselves to any possible position in which they may be moved by the action of an aeroplane while in flight.' A further optimistic claim was that the aviator could be blindfolded 'so that the sense of direction may be sensitized without the assistance of the visual senses. In this way the aviator when in fog or intense darkness may be instinctively conscious of his position'.

2.3   The beginnings of a systematic approach

To a simulation engineer, a faithful simulation requires the following three elements:

(1)   a complete model, preferably expressed mathematically, of the response of the aircraft to all inputs, from the pilot and from the environment

(2)   a means of solving these equations in 'real-time', or in other words, of animating the model

(3)   a means of presenting the output of this solution to the pilot by means of mechanical, visual and aural responses

None of these has yet been completely solved, if judged by the strictest engineering criteria – whether or not they need to be is another question.

Enough knowledge had been gained of the mechanics of flight by about 1920 to produce a reasonable mathematical model of flight (Bairstow 1920), but no means yet existed to translate a complete model into a usable simulation. The early automatic controllers produced by Sperry and others relied on an empirical understanding of aircraft behaviour and practical testing rather than simulation (Bennett 1979). However, evolution of the technology of automatic control provided the basis for improved simulation; all of the elements – sensors, actuators and computing

Fig. 2.3. Lender and Heidelberg inventions. (Courtesy Controller HMSO.)

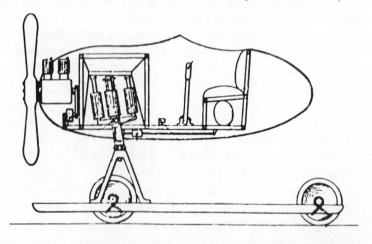

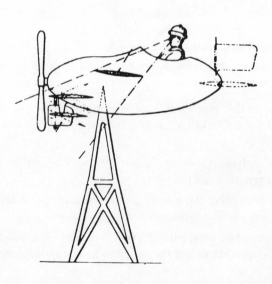

mechanisms – were common to both disciplines. As in Sperry's work, computing mechanisms could be built and adjusted relying on a practical understanding of how systems behaved. This was the approach in the first attempts at simulation.

One of these was that of Lender and Heidelberg described in patents of 1917 and 1918 (Lender & Heidelberg 1917, 1918). In their invention, the motions of a cockpit balanced on a support were produced in three axes by compressed air actuators. A new feature was a simulation of the effects of speed. Fore and aft motions of the stick were integrated pneumatically to result in the mechanical displacement of a rod, which in turn acted on the controls to alter their effect. The illustrations (Fig. 2.3) show two examples of their ideas in which a propellor is used for additional realism. A projection apparatus was outlined for producing a visual display, but only in sketchy detail.

This line of development continued, the features being a pivoted cockpit, actuators able to produce motion in three axes in response to stick and rudder, and some means of introducing disturbances. In all cases the intention was to reproduce the 'feel' of the aeroplane. Most could not approach this because of the crude or non-existent dynamic simulation. One which got nearer than most was the Link Trainer.

Edwin Link gained his early engineering experience with his father's firm, the Link Piano and Organ Company of Binghamton, New York, and in fact Link's first patent was granted for an improvement in the mechanism of player pianos (Kelly 1970). The trainer was developed in the period 1927–29 in the basement of the Link factory and made use of pneumatic mechanisms familiar from the piano and organ business. The first trainer, advertised as 'an efficient aeronautical training aid – a novel, profitable amusement device' was described in a patent filed in 1930 (Link 1930). Pitch, roll and yaw movements were initiated in the same manner as in its predecessors, but pneumatic bellows were used for actuation. An electrically driven suction pump mounted in the fixed base fed the various control valves operated by the stick and rudder, while another motor-driven device produced a repeated sequence of attitude disturbances. Link worked for a long period, adjusting the performance by trial and error, until he achieved a satisfactory feel.

The first description of the trainer (Fig. 2.4) made no reference to instruments; the device was primarily intended to demonstrate to students the effect of the controls on the attitude of the simulated aeroplane and to train them in their coordinated operation. The simulated effects of the ailerons, elevators and rudder were independent. They did not represent the interactions present in the real aeroplane. Also, because of the direct cause and effect link between the controls and the actuators, its motion served to indicate attitude rather than to provide correct motion cues.

Link had a difficult job convincing people that his device was worth the investment. It may have had a more realistic feel to it than its predecessors, but it was still not seen to meet a real training need. Such a need did exist, however, in instrument flying. This requirement demanded a more analytical approach to simulation – a model of the aircraft behaviour had to be set up first.

Every flight training device described so far has had a movable cockpit indeed this was generally the sole function of the simulator. Given that this form of motion is not useful motion cue, a simulator providing instrument displays can dispense with it altogether. This was, in fact, done in all succeeding trainers except the Link until the era of true motion cue simulation.

Rougerie's patent of 1928 (Rougerie 1928) describes a simple trainer, fixed to the ground, consisting of a student's seat facing an instrument panel and two sets of controls, one each for the student and instructor. The student's flight instruments are directly connected to the instructor's controls. The student would fly the trainer in response to commands from the instructor, who modifies the instrument readings according to the

Fig. 2.4. The first Link.

student's actions. A more automated version of this scheme can be seen in Johnson's invention of 1931 (Johnson 1931). W.E.P. Johnson, an instructor at the Central Flying School, Wittering, and one of the pioneers of instrument flying in Britain, constructed his trainer from a written-off Avro 504 fuselage (Taylor 1958). In the simplest form of this invention an airspeed indicator, turn indicator and bank indicator were directly operated by wires attached to the sticks and rudder bars of student and instructor. Further improvements included a throttle control affecting the airspeed indicator and integrating devices for the display of altitude and heading. It seems that Johnson's invention achieved its objectives but was not developed further as his flight 'was too preoccupied with real flying to develop it further'. If he had done so, it is likely that a useful simulator would have resulted.

Another early British instrument flight trainer is that described by

Fig. 2.5. Roeder's aeroplane model.

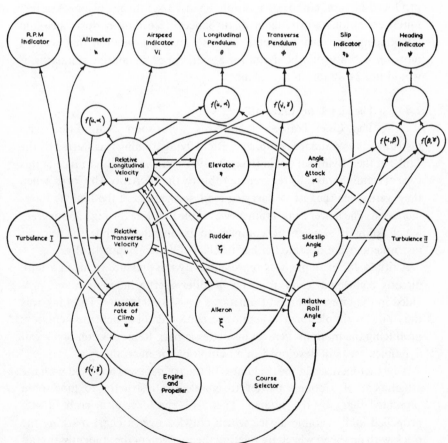

Jenkins and Berlyn of Air Service Training Limited, Hamble, in their patent applications of 1932 (Jenkins & Berlyn 1932). This ground-fixed apparatus used mechanisms similar to Johnson's for linking the instruments to the controls. Simple mechanical means were used to produce the required dynamic behaviour of the instruments in response to control inputs, but on a case by case basis rather than as an outcome of a unifying dynamic model.

The first simulator described in which the design followed the systematic approach outlined at the start of this section must be that of Roeder (Roeder 1929). On every re-reading, one comes away with the impression that he has said it all! Roeder's patent describes in detail a simulator for the height control system of an airship. The computer is part hydraulic and part mechanical, with cams for generating non-linear functions. The diagram (Fig. 2.5) reproduced in translation from the patent, shows the interaction of forces in an aeroplane together with the inputs (controls) and outputs (instruments). From a scheme such as this Roeder produced his simulation. In addition, he discusses the problems of setting initial conditions, introducing disturbances, recording performance and introducing instrument failures. He also gives the opinion that while a movable cabin would be useful for an airship or submarine simulator, it would not be so for an aeroplane.

2.4   The simulator takes off

The Link Trainers themselves were soon being fitted with instruments as standard equipment. Blind flying training was started by the Links at their flying school in the early 1930s and as the importance of this type of training was fully realised, notably by the US Army Air Corps, when they were given the task of carrying mail, so the sales of the Link Trainers increased. The newer Link Trainers were able to rotate through 360 degrees which allowed a magnetic compass to be installed, while the various instruments were operated either mechanically or pneumatically.

Altitude, for example, was represented by the pressure of air in a tank directly connected to an altimeter. Rudder/aileron interaction was provided in the more advanced trainers, as was a stall feature. The Link was thus now a simple form of analogue computer of which the actuators producing the motions were an inseparable part. Reproduction of aircraft dynamics was still developed in an empirical manner, though.

A further increase in the usefulness of the trainers was achieved with the attachment of a course plotter to enable the instructor to monitor a simulated flight by the student. This took the form of a 'crab', a self-propelled and steerable device which crawled over a chart marking the track with an inked wheel. By relating the position of the student's aircraft to marks on the chart, the instructor was able manually to control the transmission of simulated radio beacon signals to the trainer (Fig. 2.6).

The Link achieved increasing success throughout the 1930s. The device was produced in various versions and was sold to many countries, including Japan, the USSR, France and Germany. The first Link Trainer to be sold to an airline was that delivered to American Airlines in 1937. The

Fig. 2.6. Link Trainer showing instructor's table.

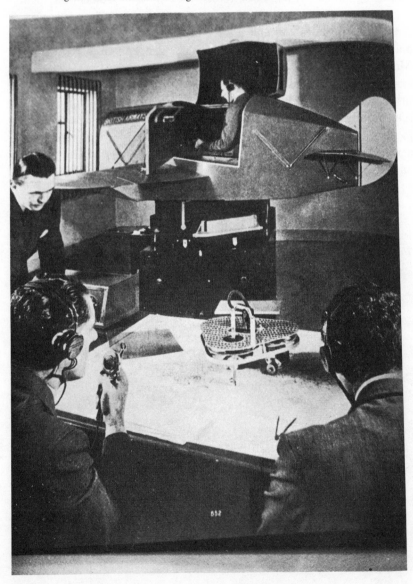

RAF also took delivery of their first Link in that year. By the beginning of the Second World War, many of the major air forces were doing their basic instrument training on Links, or derivatives. At the start of the war, German pilots posted to bomber squadrons had had 50 hours of blind flying training on Link Trainers (Deighton 1977) (Fig. 2.7).

The need also arose for the training of large numbers of recruits in the many individual and team skills involved in the operation of the ever increasing number of military aircraft types. Basic instrument instruction was performed in part on Link Trainers (Curtis 1978), but aircraft developments such as variable pitch propellors, retractable undercarriages and higher speeds made sound training in cockpit drill essential (Air Member for Training 1945). The mock-up fuselage was one solution – for example, the Hawarden Trainer, made from the centre section of a Spitfire fuselage, enabled training in the procedures of a complete operational flight (Directorate of Operational Training 1942). The Links, too, were developed to the stage where the instrument layout and performance of specific aeroplanes were duplicated. The US Army–Navy Trainer, Model 18 (ANT-18), for example, was designed for indoctrination in AT-6 and SNJ flying.

Fig. 2.7. Link Trainer with cyclorama.

In 1939 Britain requested Link to design a trainer which could be used to improve the celestial navigation capabilities of their crews who were ferrying 'surplus' US aircraft across the Atlantic. Also, it was hoped that such a trainer could be used to improve bombing accuracy during night raids over Europe (Kelly 1970). Edwin Link, together with the aerial navigation expert, P. Weems, worked out the design of a massive trainer suitable for use by an entire bomber crew, the whole to be housed in a 45

Fig. 2.8. The celestial navigation trainer.

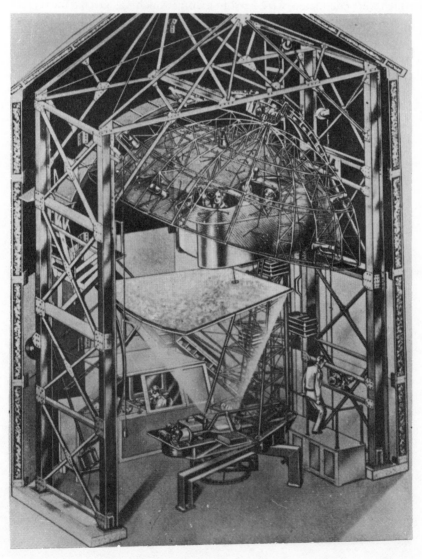

foot high silo-shaped building. This was the Celestial Navigation Trainer (Fig. 2.8).

The trainer incorporated a larger version of the conventional Link Trainer fuselage which could accommodate the pilot, navigator, and bomber. The pilot flew the trainer, which included all of the normal Link facilities and instruments, while a bomb aimer's station provided the appropriate sight and an image of targets over which the trainer flew. The navigator was provided with radio aids and, in addition, a very elaborate celestial view from which he could take his astro sights. The 'stars', twelve of which were collimated, were fixed to a dome which was given a movement to correspond with the apparent motion of the stars with time and changes in bomber longitude and lattitude. The first Celestial Navigation Trainer was completed in 1941, and the RAF placed an order for sixty. However, only a limited number were installed. Hundreds of them were installed and operated in the United States (Maisel 1944).

Throughout the war, many instructors on RAF stations contributed their ideas with the construction of improvised devices – this due to the long delivery times and low priority given to the manufacture of training aids. An early development was the 'instructional fuselage', consisting of a fuselage of the required type mounted on stands and housed in a hangar. 'It could be used to train air crews in all the drills they have to carry out in the particular aircraft that they are being trained on. All the services, hydraulic, electrical,

Fig. 2.9. A Silloth Trainer for a Halifax aircraft. (Courtesy Controller HMSO.)

and pneumatic, and their recording instruments are made to work in the normal manner, so that the various drills carried out by the crew are realistic. Bomb-dropping procedure and abandon aircraft drills by parachute and dinghy are also carried out; the bombs are released into sand trays beneath the aircraft.' (Directorate of Operational Training 1942.) Other reports exist of trainers made from fuselage sections mounted on crude motion systems.

One such trainer which achieved wider success in the RAF was the 'Silloth Trainer', developed by Gordon Iles at RAF Silloth, a Coastal Command station near Blackpool. This trainer was designed for the training of all members of the crew, and was primarily a type familiarisation trainer for learning drills and the handling of malfunctions. As well as the basic flying behaviour, all engine, electrical and hydraulic systems were simulated, and sounds generated. An instructor's panel, visible in the illustration, was provided to enable monitoring of the crew, and fault insertion (Fig. 2.9)

All computation was pneumatic – a natural choice for the designer, whose background in the pianola business gave him a lot in common with Edwin Link. The simulation was developed empirically but when properly adjusted, gave realistic responses. Silloth Trainers were manufactured for many types of 2- and 4-engined aircraft throughout the war, but were never used in large numbers; in mid-1945 only 14 were in existence or on order. Later in the war improved versions of this type of trainer were designed at other RAF stations but were never developed since all such work ceased at the end of the war.

It is interesting to note in passing the contribution of Rediffusion. The company's skills in wired radio distribution enabled them to develop crew trainers for radio navigation. Many of these were manufactured and used by the RAF (Adorian, Staynes & Bolton 1979). Rediffusion did not get into the flight simulation business until after the war.

2.5   The electronic simulator

The Silloth Trainer demonstrated that the mechanical and pneumatic techniques had reached the end of their usefulness, at least for detailed simulation of specific aircraft types. Electrical methods of analogue computation were known (Soroka 1954) but it required the urgency and cross-fertilisation of ideas of World War II for the necessary development of the technique to occur. The analogue computer, or differential analyser as it was known then, enabled a calculation of the response of the vehicle to aerodynamic forces as opposed to an empirical duplication of their effects.

In 1936 Mueller, at MIT, described an electronic analogue computer for the faster than real-time simulation of aeroplane longitudinal dynamics (Mueller 1936). His interest was in aircraft design and the solution of the equations of motion, but as a postscript to his paper he mentioned the possibility of extending the timescale of the simulation and of including a man in the loop.

Two of the first flight trainers which used an electrical form of computation were Dehmel's machine and Travis' 'Aerostructor'. Richard Dehmel, an engineer with Bell Telephone, become interested in the problems of flight training in 1938. His first development was an automatic signal controller for generating synthetic radio signals with a Link Trainer, thereby eliminating the need for an attendant to manually adjust the volume controls. Following this, Dehmel developed the 'flight' portion of a trainer based on electrical circuits (Dehmel 1941). This machine was never manufactured, but served as a starting point for future developments by the inventor.

The Aerostructor, developed by A.E. Travis and his colleagues in 1939–40, also in the USA, was a fixed base, electrically operated trainer with a visual rather than an instrument presentation. The visual system was based on a film loop and simulated the effects of heading, pitch and roll movement (Flight Training Research Association 1940). The trainer was widely exhibited, but was never commercially produced. It was however used in large numbers by the US Navy in a modified form as the 'Gunairstructor'.

In 1941 an electronic simulator which solved aircraft equations of motion was designed at the Telecommunications Research Establishment (TRE) in Britain, famous for its radar work. This simulator was designed to provide the 'flying unit' for the TRE aerial interception (AI) radar trainers (Dummer 1985). The computer was based on the ideas of F. C. Williams (later to become one of the digital computer pioneers), and used the Velodyne (Williams & Uttley 1946), another TRE invention, for integration. The dc method of computing was used in the simulation of the simplified aerodynamics of a fighter. Many of these flying units were used in TRE trainers during the war (Fig. 2.10). Later, in 1945, a more advanced unit including feel forces was designed by A. M. Uttley for use in a new AI visual crew trainer. This, however, never went into service.

In addition to using advanced electronic computational methods, the TRE AI trainers were further examples of full crew trainers. The four stages of AI combat could be practised: following an interception course provided by a ground operator, the 'chase' guided by on-board radar, the visual contact, and the moment of firing. The TRE Type 19 provided training in the complete sequence by including positions for the pilot and AI operator, an instructor's unit, computers for simulation of the attacking aircraft and the relative position of the 'enemy', a visual projection unit and a course recorder. Some of these trainers were built as mobile units whose function was to tour operational squadrons to train in the use of the latest versions of airborne radar.

In 1941 Commander Luis de Florez, of the US Navy, visited Britain and wrote his 'Report on British Synthetic Training'. This report was influential and helped to bring about the establishment of the Special Devices Division of the Bureau of Aeronautics, the forerunner of the present Naval Training Systems Center (Murray 1971). Also in that year, a copy of the Silloth Trainer was constructed in the USA for evaluation at the Mohler Organ plant. As a result, it was decided to build an electrical version, as instability of adjustment due to humidity, temperature and ageing made the system unmanageable. The task of designing this was given to the Bell Telephone Laboratories who produced an operational flight trainer for the Navy's PBM-3 aircraft. This device, completed in 1943, consisted of a PBM front-fuselage and cockpit, with complete controls, instrumentation and auxiliary equipment, together with an electronic computing device to solve the flight equations (Huff 1980). Non-linear functions were generated by contoured potentiometers driven by the integrator servos. The simulator had no motion or visual systems or variable control loading. A total of 32 of these simulators for seven types of aeroplane were built by Bell and Western

Fig. 2.10. A TRE flying unit. (Courtesy Controller HMSO.)

Electric during the war years (De Florez 1949). It has been stated that the PBM-3 was 'probably the first operational flight trainer that attempted to simulate the aerodynamic characteristics of a specific aircraft' (Dreves, Pomeroy & Voss 1971), but this is a questionable claim.

Dehmel continued to develop his ideas independently of Bell and managed to interest the Curtiss–Wright corporation in the manufacture of his devices in 1943. After the development of a prototype trainer, the US Air Force ordered two trainers from Curtiss–Wright for the AT-6 aeroplane. Production of these followed. (Fig. 2.11).

After the war, competition from Curtiss–Wright stimulated the Link Company to develop their own electronic simulators. Also at this time the value of the Link Trainer motion system was being called into question (Kelly 1970). The movements of this trainer did not correctly simulate the forces experienced in flight, and in fact a ground-fixed trainer would accurately locate the force vector in more cases. Also, the axis of roll rotation was too far below the pilot to allow correct simulation of accelerations due to roll. It was argued that the modern pilot should not fly 'by the seat of his pants', but by instruments. Ed Link disagreed with this, holding the view that trainer motion was needed even if incorrect, since motion was present in flying. Despite this, Link followed the trend to fixed base simulators. The company therefore developed their own electronic analogue computer which was used in the C-11 jet trainer (Fig. 2.12). A

Fig. 2.11. A Curtiss–Wright Z-1 without cover. (Courtesy R. C. Dehmel.)

contract was awarded by the US Air Force in 1949, and eventually over a thousand of these types were sold.

Meanwhile, Curtiss–Wright had contracted to develop a full simulator for the Boeing 377 Stratocruisers of Pan American Airways. The simulator was installed in 1948 and was the first full aircraft simulator to be owned by an airline. No motion or visual systems were installed, but in all other respects the simulator duplicated the appearance and behaviour of the Stratocruiser cockpit. The trainer was found especially useful for the practice of procedures involving the whole crew; emergency conditions could readily be introduced by the instructor on his comprehensive fault insertion panel. Complete routes could be flown, as in real life, using the same navigational aids. This facility was used by other airlines, and impressed users by the degree of procedural realism achieved (Brice 1951). The lack of motion, though, did give rise to reservations due to the unnatural feel of the aeroplane, even giving rise to controlling problems.

In Britain, a similar simulator was built for BOAC by Rediffusion under license to Curtiss–Wright. Rediffusion later built a simulator for the Comet I for the same airline.

The first Curtiss–Wright, Rediffusion and Link simulators used the ac carrier method of analogue computing. Air Trainers Ltd (the successor

Fig. 2.12. A Link C-11.

company to the first importers of Link Trainers to Britain, and later to be merged with Rediffusion) however, decided to use the dc method, a more demanding but potentially more precise technique. Their first simulator using this method was built for the RAF's Meteor aeroplane. Further developments in ac simulation were made, and of course the electronics technology moved from valves to transistors, enabling smaller and cooler running analogue computers.

2.6   The modern simulator takes form

In the early 1950s aircraft manufacturers did not have a great deal of analytical information on the performance of their airframes and engines – the gaps had to be filled by the simulator manufacturers by trial and error and pilot evaluation. This situation changed, though, with more and more data becoming available from flight testing programmes. Together with requirements for driving the motion and visual systems then being introduced and pressure from the operators for improved accuracy (and, they hoped, better transfer of training) significant increases in the amount of analogue computer hardware became necessary. The law of diminishing returns started to operate and the cumulative errors of all of this interacting circuitry exceeded the improved accuracy implied by the data available. Another result observed was a decrease in reliability despite improved component technology. The effect of the latter was multiplied by the increasing utilisation being made of simulators.

This period coincided with the introduction of the second generation of digital computers. These machines had the potential for solving the accuracy and reliability problems, and at a cost which was now low enough to be practical for most applications. As a consequence, there was an almost complete shift to digital simulators, with analogue computers only being retained for the simplest trainers and those parts of a simulation where high enough performance could not be achieved digitally at reasonable cost.

The idea of using numerical techniques in control systems was suggested as early as 1942, the year in which P. Crawford submitted his masters thesis 'Automatic Control by Arithmetical Operations', to MIT (Redmond & Smith 1980). Crawford's later influence contributed to the definition of the first digital flight simulator project, again started at MIT. This project had its origins in the US Navy's desire for a universal flight simulation machine which could be used for both aircraft development and training. A research contract was awarded to the Servomechanisms Laboratory at MIT in 1943 to develop ASCA (Airplane Stability and Control Analyser). This project was started on conventional analogue computer lines, but due to the influence of Crawford's work and the other digital computer projects underway elsewhere, ASCA too became a digital project. It was realised that the powerful computer being planned would be a general purpose one, and wider applications were found. The machine became the Whirlwind computer. The requirement for a flight simulator had led to its specification, but it in fact never became one.

The US Navy also funded research at the University of Pennsylvania (home of ENIAC) where the UDOFT (Universal Digital Operational Flight Trainer) project started in 1950. This computer was manufactured by the Sylvania Corporation and completed in 1960. The UDOFT project demonstrated the feasibility of digital simulation and was mainly concerned with the solution of the aircraft equations of motion.

In the early 1960s Link developed a special purpose digital computer, the Link Mark I, designed for real-time simulation. This machine had three parallel processors for arithmetic, function generation and radio station selection. In the late 1960s, general purpose digital computers designed primarily for process control application were found also to be suitable for simulation, with its large real-time input/output requirement, and thus the use of special purpose machines declined. Today special purpose digital computers are only used in applications demanding very high speed processing, such as radar and image simulation.

Nearly all of the simulators produced up to the mid 1950s had no fuselage motion systems. This could be justified by stating that it was procedures that were being taught, but the fact remained that fixed-base simulators did not 'fly' like aeroplanes. It was found that handling improvements could be made by trial and error adjustments of the control loading and flight dynamics simulations which partly compensated for the lack of motion (Cutler 1966). Proposals were made by the manufacturers for motion systems, but it was not until the late 1950s that the airlines decided to purchase them. In 1958, Rediffusion produced a pitch motion system for a Comet IV simulator (Fig. 2.13). More complex motion systems were designed capable of producing accelerations in up to six degrees of freedom. A detailed review of the history of motion simulation is given by Clark (Clark 1962).

Systems for producing the external visual scene have been proposed and constructed for almost as long as flight simulators. However, realistic and flexible visual attachments are a fairly recent development. Due to the enormous number of visual systems which have been invented, the majority unsuccessful, only a few can be mentioned here. The point light source projection, or shadowgraph, method enjoyed popularity in the 1950s, especially for helicopter simulators. A series of simulators using this method of display was produced by Giravions Dorand in France (Hill & de Guillen-

Fig. 2.13. Comet IV with pitch motion. (Courtesy Rediffusion Ltd.)

Fig. 2.14. Shorts helicopter simulator. (Courtesy Short Brothers Ltd.)

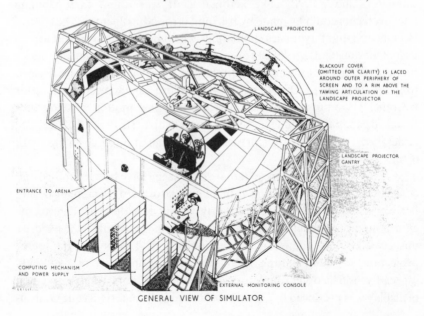

schmidt 1955), including an ab initio hovering trainer. A similar system built by Shorts is illustrated (Fig. 2.14). Simulators on this pattern were also built in the USA, but the shortcomings of the shadowgraph system seem to have limited the success of the idea. The first visual system achieving widespread use on civil aviation simulators were based on a scale model viewed through a television system, although methods based on film viewed through anamorphic optics have also met with success in more restricted applications. Serious development of closed circuit television visual systems began in the mid 1950s with monochrome versions being produced by all of the major manufacturers. Rediffusion produced the first colour system in 1962.

The first computer image generation systems for simulation were produced by the US General Electric Company for the space programme. Early versions of these systems produced a patterned 'ground plane' image, while later ones were able to generate images of three-dimensional objects (Elson 1967). Progress in this technology has been rapid as its performance is closely linked to advances in microelectronics technology. The first computer image generation systems economically feasible for commercial simulator operators were of the 'night only' type. These used the calligraphic or stroke-writing, rather than the scanned raster method of display. This enabled a simple and superior simulation of light points. The first of these, known as Vital II, was produced by the McDonnell–Douglas Electronics Corporation in 1971.

To continue the history at this point would be to encroach on the following chapters. The flight simulator arrived at its modern form probably at the end of the 1960s. Improvements made since then have largely been refinements, some very significant, to the basic principles established by that time. Most of these had their origin during the Second World War period, the watershed of simulator development. Before that period, simulators were the playthings of amateur engineers – after it the discipline had become professionalised and industrialised, with the role of the individual inventor becoming a secondary one.

3

Mathematical models for flight simulation

3.1 Introduction

3.1.1 *Basic concepts*

This chapter deals with mathematical models for flight simulation. The principal task, that of modelling the dynamic behaviour of the flight vehicle, is an extension of the basic principles of flight mechanics outlined in such well-known texts as Babister (1961) and Etkin (1959). Simulation must deal with more than just airframe aerodynamics, however, and broader aspects of mathematical modelling will be discussed.

The nature of the problem is that the motion of a flight vehicle is governed by equations of motion of the generic form

$$\ddot{x} = F/m \tag{3.1}$$

where $\ddot{x}$ is the acceleration of the vehicle, F is the applied force and m is the mass.

The specific form of the equations of motion must be established, and this will be dealt with later in Section 3.3, but given an expression like equation (3.1), then the basic mathematical model of the vehicle is embodied in the definition of F. For a vehicle flying in air, this mathematical model is primarily the relationship between the air reactions and the motion of the aeroplane relative to the air. This can be called the aerodynamic model. Other external forces and moments arise from engine thrust and landing gear ground contact. Definition of all these force and moment components is the key to a realistic description of an aircraft's flight characteristics. An aircraft's performance and dynamic behaviour can then be calculated for as wide a range of flight conditions as that for which F can be defined. Simulation is fundamentally the generation of these forces and the solution of the equations of motion.

The term 'mathematical model' is thus growing in complexity and extent, and is also applied to other features of the aircraft such as its control system, and many internal systems. For each feature being modelled, it is necessary to formulate the model so that the 'behaviour' of the system in some sense, i.e. the response of the system to a stimulus, may be calculated. A more detailed discussion will be given in the next Section.

A model is only useful if it can actually be used for some application. So far, the term model has not properly separated the concept from the implementation. This may usefully be done, with the aid of some definitions outlined by Schlesinger (1979).

The simulation space is divided into three basic elements as depicted in Fig. 3.1. Starting with Reality, a conceptual model, described via equations or other governing relationships, is obtained by analysis. Implementation via computer programming yields a Computer Model which, through simulation, may be related to Reality. The credibility of the conceptual model is then evaluated by procedures which test the adequacy of the model to provide an acceptable level of agreement with reality, while the computer model, in the form of an operational computer program, is confirmed as an adequate representation of the conceptual model by procedures of verification. Finally, model validation demonstrates that the computer model possesses a satisfactory range of accuracy in comparison with reality and consistent with its intended application.

In some aircraft simulation applications, the mathematical models may be defined by one agency, such as the aircraft manufacturer, for implementation by another, a simulator manufacturer. A third party, such as a regulatory authority, may require that the original model be implemented

Fig. 3.1. The simulation space and the definition of 'models'.

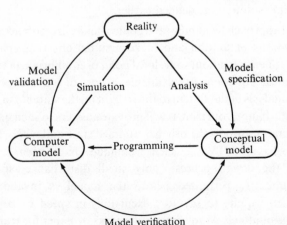

unchanged. Verification would be performed by the simulator manufacturer using his own internally-generated checks, and validation by comparison with real-life data recorded from flight tests.

In creating or deriving mathematical models, it is important that the modeller (the person doing the modelling) has a clear idea of what the model is for, and that he states this together with his definition of his model. It is important because the purpose of the model influences its form and quality (Thomas 1984). Many systems are strictly governed by equations which may be extremely complex but which may often be simplified in the interests of obtaining practical solutions and yet still retain sufficient realism for the task at hand.

A practical solution in the present context means one of adequate accuracy which can be achieved in real-time. Real-time here refers to a solution in which the calculation of a system's behaviour over, say, one second of elapsed time can be achieved in one second or less of computing time. Adequate accuracy means that the real-time solution yields the steady-state performance of the vehicle and its transient behaviour with an accuracy which is acceptable and sufficient for the role of the simulation.

3.1.2   *Mathematical modelling versus stability and control analysis*

Deriving mathematical models for simulation has some similarities with the basic techniques of stability and control theory (Babister 1961; Bramwell 1976; Etkin 1959) but there are also some differences. Similarities include mathematical notation, systems of axes and basic nomenclature. Differences, however, are significant, and include the need for:

a wide range of speed, aircraft configuration and manoeuvre

real-time solution, as defined earlier

The classical approach to stability and control analysis is to start with the complete equations of motion and make assumptions that enable the equations to be linearised about some local point of equilibrium (a trimmed state). Once linear equations are available, a host of techniques can be applied to the analysis of the system to throw light on the stability of motion following a disturbance from trim. It will provide answers to such questions as 'will the aircraft return to the original trimmed state and with what sort of transient behaviour?' While such techniques have been, and are, invaluable in the design process, only small disturbances from the equilibrium state are permitted before the model is invalid. Small disturbances here equate to say 10% excursions in speed or perhaps 5 degrees in angle of attack. Many simulation tasks, whether for training or for research and development, demand a wide range of flight speed or require large changes in aircraft configuration in order to accomplish that task. A complete sortie, from take-off to landing, is an extreme example; a transition from wing-borne to jet-borne flight of a VSTOL vehicle a less extreme but still quite complex one.

Real-time solution is necessary because, in simulation, there is usually a man in the control loop, who must be supplied with information in a timely fashion so that he can use his normal control techniques and strategies. Alternatively, flight hardware may need to be tested.

3.2   Elements of the model

The first section of this chapter introduced the concept of a mathematical model, where the primary, or core elements were defined as those which directly produce the external forces acting on the airframe, namely aerodynamics, engine thrust and undercarriage.

Secondary elements of a model contribute significantly to the core, and include a control system and an atmospheric environment.

All these features are summarised in Fig. 3.2 and are discussed below.

The aerodynamic model has to reproduce the dominant features of the forces and moments acting on the aircraft: lift and drag, the main contributors to the aircraft's performance, and moments about all three axes, through which the pilot exercises control. Mathematical models would normally consider the physical components of the vehicle, e.g., wing, body, tail or, for a helicopter (Bramwell 1976), main rotor, tail rotor, body and tail-plane, and build mathematical expressions for the contributions made by these components. Section 3.4 outlines this process in more detail. Additional features may need to be included to represent such influences as

Fig. 3.2. Contributions to the aircraft mathematical model.

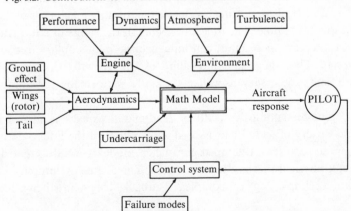

'ground effect', whereby the presence of the ground close to the aircraft during take-off and landing so constrains the airflow round the aircraft that the forces acting differ markedly from those applicable to the same configuration in free air. This can result in increased wing lift and possibly in a sharp increase in nose-down pitching moment close to the ground. Both of these effects are important as they influence the pilot's ability to perform a safe and smooth landing. Aerodynamic models are normally static or quasi-static, based on the assumption that the airflow can re-establish steady conditions in a time scale much shorter than that of the aircraft manoeuvre itself. Thus time, as such, does not appear explicitly in the mathematical model, but the aircraft's transient behaviour is, of course, a function of time.

An engine model must first produce the correct value of steady thrust to correspond with the pilot's demand through his power lever. Thrust can be modelled as net thrust i.e. the net propulsive force, having allowed for intake momentum drag. A more comprehensive model will reproduce gross thrust and momentum drag as separate entities, taking the difference explicitly rather than implicitly. This may be particularly important in cases where the aircraft is capable of flying at high angles of attack or can vector its engine exhaust nozzles in some way, as the thrust vector and momentum drag vector may then deviate by a large angle. Interference between the engine exhaust flow and the tailplane may further affect the available applied forces; thrust vectoring being a major influence here. An additional example of the need for detailed engine modelling can occur with VSTOL aircraft which, in the hover, must use a proportion of the airflow through the engine to generate control forces by blowing out at the extremities. An over-active pilot may be disconcerted to find that he descends as a result of his control activity.

An engine can not change its thrust instantaneously. Burning more fuel to increase thrust must first accelerate the rotating machinery of the engine and this takes time. The timescale, being in the range of a few tenths of a second to several seconds, can influence the pilot's ability to control the aircraft. Thus, adequate modelling of the growth or decay of thrust with time is necessary, as well as the static, performance-orientated thrust modelling.

Engine modelling, especially of the dynamic aspects, is an area where a wide variety of techniques is used, depending on the fidelity in dynamic response required. A 'lumped parameter' model is usually created, rather than a physical thermodynamic model. The degree of 'lumping' can vary from a simple lag to complex feedback loops involving fuel flow, temperature limiters, rpm governors and the like. A fuller description of engine modelling is given in Chapter 4.

If a simulated aircraft is going to take-off or land, it must have some means of support when in contact with the ground. An undercarriage model provides this support. A real undercarriage is a complex mechanical, hydraulic and sometimes pneumatic assembly, which can be modelled in a variety of ways. Providing some ground reactions sufficient to balance the aircraft's weight can be done by a simple spring and damper model, with extra damping on recoil. Depending on the simulation task, however, a detailed model may be necessary, to include tyres and brakes, anti-skid systems and nose-wheel steering.

The discussion so far has been concerned with the contributions to the external forces. Thus

$$F = F_{aero} + F_{engine} + F_{undercarriage}$$

Referring back to the generic equation (3.1) given in the Introduction, the other item which also needs modelling is the vehicle's mass or, more generally, its mass properties. These comprise its mass, its moments of inertia and its centre of gravity location, usually all as a function of payload, fuel state and geometry, e.g., wing sweep, or stores configuration.

A full atmospheric model needs to represent a number of properties. The starting point is the variation of air density and temperature with altitude, as defined by the International Standard Atmosphere (ESDU 1968). Whether for design checks, or training purposes, non-standard ambient conditions such as a 'hot' day (e.g., ISA + 10), may also need to be simulated. These properties may be called atmospheric statics. A wind environment is also needed, in terms of mean surface wind for take-off and landing, and winds at altitude as they affect performance on a route schedule, for example. Associated with wind, there is usually turbulence, the effect of which can range from a distraction and disturbance in a precise task, to a design case for control authority or for crew ride comfort. Methods of modelling turbulence have improved in recent years (Jansen 1982; Tomlinson, 1975) as a result of active research into the real world environment. More extreme atmospheric phenomena are also now being modelled. Wind shear has caused several major accidents and is a legitimate item to be simulated, either for research into aircraft design or for training pilots to recognise and counter it effectively.

Many training tasks demand a full range of meteorological phenomena to be simulated, including weather radar returns and runway state (Klehr 1983, 1984).

The control system also needs modelling. Interposed between the pilot's control stick and the vehicle's control surfaces, it consists typically of a system of levers, cables and bell-cranks, terminating in a hydraulically-actuated jack. Various mechanical deficiencies may need to be modelled, such as cable stretch, control jams, etc. Current military aircraft and projected civil transport aircraft are replacing mechanical connections by electrical ones, usually with computers included to shape the pilot's demands and to process feedback signals. While, in some senses, these developments simplify the modelling task by removing the need to represent often inadequately known mechanical systems, the modelling needs are growing dramatically as more and more functions are included in the flight control computer. Incorporating real flight hardware in the simulator may be the best solution.

3.3   Equations of motion

Following the general discussion of mathematical models given above, it is now appropriate to consider some explicit mathematical formulations. Equations of motion will be outlined in this section, with some forms of representation for data in Section 3.4.

3.3.1   *The general equations*

In the exposition that follows, it is assumed that the reader is familiar with the simpler concepts of aerodynamics and the basic theory of aircraft stability and control (Babister 1961; Bramwell 1976; Etkin 1959; Hopkin 1966).

The mathematical formulation used here adopts the familiar notation of alphabetical triads to represent the three components of force, moment, linear and angular velocity. These are illustrated in Fig. 3.3 for the

Fig. 3.3. Axes and notation.

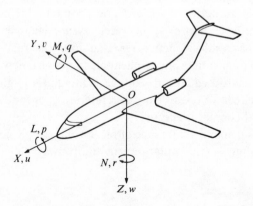

conventional right-handed orthogonal axes, with origin O fixed in the aircraft. No particular set of axes is yet assumed.

The equations of motion of a rigid aircraft flying in still air are, when referred to any system of axes fixed in the aircraft and rotating with it

$$
\left.\begin{aligned}
m(\dot{u} + qw - rv) &= X + mg_x \\
m(\dot{v} + ru - pw) &= Y + mg_y \\
m(\dot{w} + pv - qu) &= Z + mg_z
\end{aligned}\right\} \tag{3.2a}
$$

$$
\left.\begin{aligned}
I_x\dot{p} - I_{yz}(q^2 - r^2) - I_{zx}(\dot{r} + pq) - I_{xy}(\dot{q} - rp) - (I_y - I_z)qr &= L \\
I_y\dot{q} - I_{zx}(r^2 - p^2) - I_{xy}(\dot{p} + qr) - I_{yz}(\dot{r} - pq) - (I_z - I_x)rp &= M \\
I_z\dot{r} - I_{xy}(p^2 - q^2) - I_{yz}(\dot{q} + rp) - I_{zx}(\dot{p} - qr) - (I_x - I_y)pq &= N
\end{aligned}\right\}
$$

$$\tag{3.2b}$$

These are sometimes known as the 'total force' equations, as opposed to the small perturbation equations of classical stability and control.

The first three equations (3.2a) are the force equations and the second three (3.2b) are the moment equations in their 'complete' form. The full non-linear set of equations is sufficient to cope with general, large-scale motion of an aircraft, but is not soluble in this form except by numerical techniques, and then only once the forcing terms X, Y, Z, L, M, N on the right-hand side can be prescribed.

The moment equations (3.2b) are 'complete' in the sense that all the cross-inertia terms have been retained. If it can be assumed that the aircraft has a symmetrical distribution of mass with respect to the fore-and-aft plane of symmetry, then $I_{yz} = I_{xy} = 0$ and the equations become a little more compact.

$$
\left.\begin{aligned}
I_x\dot{p} - I_{zx}(\dot{r} + pq) - (I_y - I_z)qr &= L \\
I_y\dot{q} - I_{zx}(r^2 - p^2) - (I_z - I_x)rp &= M \\
I_z\dot{r} - I_{zx}(\dot{p} - qr) - (I_x - I_y)pq &= N
\end{aligned}\right\} \tag{3.3}
$$

Such an assumption is reasonable for conventional aircraft, such as a transport or executive aircraft, but a fast-jet aircraft could well have, either intentionally or as the result of a failure, an asymmetric distribution of stores beneath its wings, in which case the complete equations (3.2b) are necessary. Indeed, the inertia distribution can have a major influence on the flying qualities of a fast-jet aircraft and must be included with care.

While simple in appearance, the force equations (3.2a) defined above actually pose some practical difficulties in their solution. These difficulties are associated firstly with the product terms qw, etc. which, although nominally second-order, can become dominant in large-scale, vigorous manoeuvres, and secondly with the 'still air' assumption. Some organisations, e.g. NASA (McFarland 1975) and RAE (Tomlinson 1979), prefer to evaluate aircraft accelerations in an earth-based frame of reference, to eliminate both difficulties. In an inertial frame, the product terms disappear, as the axis system is not rotating, and the treatment of winds is quite straightforward (Tomlinson 1979).

3.3.2   *Small perturbation equations*

In many simulation situations, there is no alternative to use of the total force equations and fortunately modern computer techniques now permit this. It may still be desirable, however, to seek a linearised form of the equations, since they then become amenable to treatment by a wide range of analytical techniques. Such procedures are important at the design stage of a new aircraft and also in research, when experiments are conducted to establish data on the fundamental handling qualities of aircraft based on simplified situations and models.

If the aircraft's motion deviates to only a small extent from a datum, trimmed state, then the well-known small perturbation equations of motion are applicable. Assuming level flight ($\theta_e = 0$), choosing to use aerodynamic-body axes (such that $w_e = 0$), and omitting some derivatives that are commonly neglected, these equations may be represented in the following form (see, e.g., ESDU 1966).

$$\left.\begin{aligned}
(D + x_u)u' \quad\quad + x_w w' \quad\quad\quad\quad\quad + g\theta' \quad\quad + x_\eta\eta' &= 0 \\
z_u u' \quad + (D + z_w)w' \quad\quad\quad - V_e q' \quad\quad\quad\quad + z_\eta\eta' &= 0 \\
m_u u' + (m_{\dot{w}}D + m_w)w' + (D + m_q)q \quad\quad\quad\quad + m_\eta\eta' &= 0 \\
- q' \quad + D\theta' \quad\quad\quad\quad &= 0
\end{aligned}\right\}$$

$$(3.4)$$

$$\left.\begin{aligned}
(D + y_v)v' \quad\quad + y_p p' \quad + (y_r + V_e)r' \quad - g\phi' \quad + y_\xi\xi' + y_\zeta\zeta' &= 0 \\
l_v v' \quad + (D + l_p)p' \quad + (e_x D + l_r)r' \quad\quad\quad\quad + l_\xi\xi' + l_\zeta\zeta' &= 0 \\
n_v v' + (e_z D + n_p)p' \quad + (D + n_r)r' \quad\quad\quad\quad + n_\xi\xi' + n_\zeta\zeta' &= 0 \\
- p' \quad\quad\quad\quad\quad + D\phi' \quad\quad\quad &= 0
\end{aligned}\right\}$$

$$(3.5)$$

The coefficients, or aerodynamic stability derivatives, are constants for a given aircraft configuration and datum flight condition, and are shown here in 'concise' form for convenience, viz.

$$x_u = -X_u/m, \; l_v = -L_v/I_x$$

Such equations are rarely used in full-blown simulations, but have relevance to research studies of handling qualities or to preliminary studies of a new design, when fundamental properties of an aircraft's dynamic behaviour may be examined. For a more complete exposition, see ESDU (1966).

3.3.3   *Aircraft orientation*

An aircraft's orientation in space is defined by its attitude, which may be specified in a number of ways. The most common method is to define a sequence of three angles, known as the Euler attitude angles.

Then, starting with a set of axes, with origin O fixed in the aircraft, initially aligned with the earth reference axes $Ox_0 y_0 z_0$ as datum, the first set of axes are brought into alignment with the body-fixed axes $Oxyz$ by rotating the first set successively through each angle in turn. The usual trio of angles consists of the heading or azimuth angle ψ, the inclination or pitch angle θ and the bank angle ϕ.

The sequence of rotations is illustrated in Fig. 3.4, with the axes sets labelled successively $x_0 y_0 z_0$, $x_1 y_1 z_1$, $x_2 y_2 z_2$, xyz. The sequence is significant, as a different order from that defined here will produce a different final orientation.

Alternative Euler angle sequences are possible, and can be found in detail in Hopkin (1966).

With the definitions given here, the attitude angles can take the following range of values

$$\psi \quad \pm \pi$$
$$\theta \quad \pm \pi/2$$
$$\phi \quad \pm \pi$$

Fig. 3.4. Aircraft orientation: the Euler attitude angles and rotation sequence.

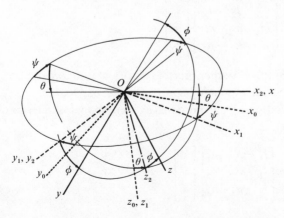

3.3.4   *Kinematic relationships for Euler angles from body rates*

Given these definitions of Euler attitude angles, derivation of the angles in practice requires a relationship between the rates of change of the attitude angles and the components of angular velocity of the aircraft's body axes (ESDU 1976b).

$$\dot{\phi} = p + q \sin \phi \tan \theta + r \cos \phi \tan \theta$$
$$\dot{\theta} = \quad q \cos \phi \quad\quad - r \sin \phi \quad\quad\quad\quad\quad (3.6)$$
$$\dot{\psi} = \quad q \sin \phi \sec \theta + r \cos \phi \sec \theta$$

and the inverse is

$$p = \dot{\phi} \quad\quad\quad - \dot{\psi} \sin \theta$$
$$q = \quad \dot{\theta} \cos \phi + \dot{\psi} \sin \phi \cos \theta \quad\quad\quad\quad (3.7)$$
$$r = -\dot{\theta} \sin \phi + \dot{\psi} \cos \phi \cos \theta$$

Note that $\dot{\phi} = p$ only when $\theta = 0$ and $\dot{\theta} = q$ only when $\phi = 0$.

Equations (3.6), also known as the gimbal equations, are widely used but do contain a singularity for, when the x-axis is vertical ($\theta = \pm 90°$), $\tan \theta = \pm \infty$ and the expressions for $\dot{\phi}$ and $\dot{\psi}$ become indeterminate.

The situations in which this occurs are clear. If such manoeuvres are avoided, then the equations may be used without difficulty. For the simulation of aircraft which perform aerobatics or similar gross ma- noeuvres, an alternative formulation is required so that the derivation of the aircraft attitude angles is always trouble-free.

3.3.5   *The direction cosine matrix*

Transformation of variables between pairs of axes is often required. One example is the derivation of the velocity components in body axes from the components in earth axes, or vice versa.

A convenient way to express this is in terms of direction cosines, viz.

$$\begin{bmatrix} x \\ y \\ z \end{bmatrix} = \begin{bmatrix} l_1 & l_2 & l_3 \\ m_1 & m_2 & m_3 \\ n_1 & n_2 & n_3 \end{bmatrix} \begin{bmatrix} x_0 \\ y_0 \\ z_0 \end{bmatrix} \quad\quad\quad\quad (3.8)$$

x, y, z represent three components (of velocity, say, or force) in body axes, and x_0, y_0, z_0 the corresponding components in earth axes.

The inverse relationship, giving earth axes components in terms of body axes components, is

$$\begin{bmatrix} x_0 \\ y_0 \\ z_0 \end{bmatrix} = \begin{bmatrix} l_1 & m_1 & n_1 \\ l_2 & m_2 & n_2 \\ l_3 & m_3 & n_3 \end{bmatrix} \begin{bmatrix} x \\ y \\ z \end{bmatrix} \quad\quad\quad\quad (3.9)$$

l_1, l_2 etc. are the direction cosines, which are defined in terms of the Euler attitude angles by

$$
\begin{aligned}
l_1 &= \cos\theta\cos\psi \\
l_2 &= \cos\theta\sin\psi \\
l_3 &= -\sin\theta \\
m_1 &= \sin\phi\sin\theta\cos\psi - \cos\phi\sin\psi \\
m_2 &= \sin\phi\sin\theta\sin\psi + \cos\phi\cos\psi \\
m_3 &= \sin\phi\cos\theta \\
n_1 &= \cos\phi\sin\theta\cos\psi + \sin\phi\sin\psi \\
n_2 &= \cos\phi\sin\theta\sin\psi - \sin\phi\cos\psi \\
n_3 &= \cos\phi\cos\theta
\end{aligned}
\tag{3.10}
$$

The direction cosines may be defined and derived in other ways, but the transformation matrices defined above are still needed and used.

3.3.6 The 'four parameter' or quaternion method

A practical alternative to the use of the Euler equations for defining the orientation of an aircraft, which avoids singularities when pitch attitude reaches 90 degrees, is to use a 'four parameter' method (Robinson 1958; Mitchell & Rogers 1965).

It is well known that one frame of axes $Oxyz$ may be brought into coincidence with a reference frame by a single rotation D about a fixed axis in space, making angles A, B, C with the reference frame $Ox_0y_0z_0$. These four parameters A, B, C, D can thus define the orientation of the axes $Oxyz$. It can be shown that the transformation matrix then relating (x, y, z) to (x_0, y_0, z_0) is given by

$$
\begin{bmatrix} x \\ y \\ z \end{bmatrix} =
\begin{bmatrix}
1 - 2\sin^2 A \sin^2\tfrac{1}{2}D & 2\cos A\cos B\sin^2\tfrac{1}{2}D + 2\cos C\cos\tfrac{1}{2}D\sin\tfrac{1}{2}D & 2\cos A\cos C\sin^2\tfrac{1}{2}D - 2\cos B\sin\tfrac{1}{2}D\cos\tfrac{1}{2}D \\
2\cos A\cos B\sin^2\tfrac{1}{2}D - 2\cos C\sin\tfrac{1}{2}D\cos\tfrac{1}{2}D & 1 - 2\sin^2\tfrac{1}{2}D\sin^2 B & 2\cos B\cos C\sin^2\tfrac{1}{2}D + 2\cos A\sin\tfrac{1}{2}D\cos\tfrac{1}{2}D \\
2\cos A\cos C\sin^2\tfrac{1}{2}D + 2\cos B\sin\tfrac{1}{2}D\cos\tfrac{1}{2}D & 2\cos B\cos C\sin^2\tfrac{1}{2}D - 2\cos A\cos\tfrac{1}{2}D\sin\tfrac{1}{2}D & 1 - 2\sin^2 C\sin^2\tfrac{1}{2}D
\end{bmatrix}
\begin{bmatrix} x_0 \\ y_0 \\ z_0 \end{bmatrix}
\tag{3.11}
$$

This can be much simplified by a change of variables from the original four parameters to a set of quaternion parameters

$$
\begin{aligned}
e_0 &= \cos\tfrac{1}{2}D \\
e_1 &= \cos A\sin\tfrac{1}{2}D \\
e_2 &= \cos B\sin\tfrac{1}{2}D \\
e_3 &= \cos C\sin\tfrac{1}{2}D
\end{aligned}
\tag{3.12}
$$

After substitution of these relationships, and use of the constraint equation

$$e_0^2 + e_1^2 + e_2^2 + e_3^2 = 1 \tag{3.13}$$

the transformation becomes

$$\begin{bmatrix} x \\ y \\ z \end{bmatrix} = \begin{bmatrix} e_0^2 + e_1^2 - e_2^2 - e_3^2 & 2(e_1 e_2 + e_0 e_3) & 2(e_1 e_3 - e_0 e_2) \\ 2(e_1 e_2 - e_0 e_3) & e_0^2 - e_1^2 + e_2^2 - e_3^2 & 2(e_2 e_3 + e_0 e_1) \\ 2(e_0 e_2 + e_1 e_3) & 2(e_2 e_3 - e_0 e_1) & e_0^2 - e_1^2 - e_2^2 + e_3^2 \end{bmatrix} \begin{bmatrix} x_0 \\ y_0 \\ z_0 \end{bmatrix} \tag{3.14}$$

This notation follows that of Mitchell and Rogers (1965) but there are alternative definitions. Comparing this matrix with the similar matrix established earlier (3.8) in terms of direction cosines, we see that

$$\left. \begin{aligned} l_1 &= e_0^2 + e_1^2 - e_2^2 - e_3^2 \\ l_2 &= 2(e_1 e_2 + e_0 e_3) \\ l_3 &= 2(e_1 e_3 - e_0 e_2) \\ m_1 &= 2(e_1 e_2 - e_0 e_3) \\ m_2 &= e_0^2 - e_1^2 + e_2^2 - e_3^2 \\ m_3 &= 2(e_2 e_3 + e_0 e_1) \\ n_1 &= 2(e_0 e_2 + e_1 e_3) \\ n_2 &= 2(e_2 e_3 - e_0 e_1) \\ n_3 &= e_0^2 - e_1^2 - e_2^2 + e_3^2 \end{aligned} \right\} \tag{3.15}$$

Furthermore, it can be shown that

$$\left. \begin{aligned} \dot{e}_0 &= -\tfrac{1}{2}(e_1 p + e_2 q + e_3 r) \\ \dot{e}_1 &= \tfrac{1}{2}(e_0 p + e_2 r - e_3 q) \\ \dot{e}_2 &= \tfrac{1}{2}(e_0 q + e_3 p - e_1 r) \\ \dot{e}_3 &= \tfrac{1}{2}(e_0 r + e_1 q - e_2 p) \end{aligned} \right\} \tag{3.16}$$

These equations provide the means to generate e_0, e_1, e_2, e_3, from the body axis components of angular velocity p, q, r. However, bearing in mind the constraint relationship (3.13) above, one of the values obtained by integration of (3.16) is effectively redundant.

Mechanisation of these equations for computer solution should recognise this constraint. One way to do this is to put (Fang & Zimmerman 1969)

$$\left.\begin{aligned}
\dot{e}_0 &= -\tfrac{1}{2}(e_1 p + e_2 q + e_3 r) + k\lambda e_0 \\
\dot{e}_1 &= \tfrac{1}{2}(e_0 p + e_2 r - e_3 q) + k\lambda e_1 \\
\dot{e}_2 &= \tfrac{1}{2}(e_0 q + e_3 p - e_1 r) + k\lambda e_2 \\
\dot{e}_3 &= \tfrac{1}{2}(e_0 r + e_1 q - e_2 p) + k\lambda e_3
\end{aligned}\right\}$$

(3.17)

where k is a constant (chosen such that $kh \leqslant 1$ for an integration step size h) and

$$\lambda = 1 - (e_0^2 + e_1^2 + e_2^2 + e_3^2)$$

(3.18)

This is the method of algebraic constraint.

Finally, expressed in terms of the three Euler angles ψ, θ, ϕ, the four parameters e_0, e_1, e_2, e_3 are

$$\left.\begin{aligned}
e_0 &= \cos\tfrac{1}{2}\psi \cos\tfrac{1}{2}\theta \cos\tfrac{1}{2}\phi + \sin\tfrac{1}{2}\psi \sin\tfrac{1}{2}\theta \sin\tfrac{1}{2}\phi \\
e_1 &= \cos\tfrac{1}{2}\psi \cos\tfrac{1}{2}\theta \sin\tfrac{1}{2}\phi - \sin\tfrac{1}{2}\psi \sin\tfrac{1}{2}\theta \cos\tfrac{1}{2}\phi \\
e_2 &= \cos\tfrac{1}{2}\psi \sin\tfrac{1}{2}\theta \cos\tfrac{1}{2}\phi + \sin\tfrac{1}{2}\psi \cos\tfrac{1}{2}\theta \sin\tfrac{1}{2}\phi \\
e_3 &= -\cos\tfrac{1}{2}\psi \sin\tfrac{1}{2}\theta \sin\tfrac{1}{2}\phi + \sin\tfrac{1}{2}\psi \cos\tfrac{1}{2}\theta \cos\tfrac{1}{2}\phi
\end{aligned}\right\}$$

(3.19)

These expressions are necessary to derive initial values for e_0 etc. when ψ, θ and ϕ are known.

In the context of aircraft simulation, the direction cosines are still required for the transformation of variables from body axes to earth axes (and vice versa), and equations (3.15) may be used to derive them from the quaternion parameters. The Euler angles themselves are also required for display on the pilot's instruments and elsewhere. A suitable selection from equations (3.10) and noting (3.15) enables the angles to be derived. Thus,

$$\sin\theta = -l_3 = 2(e_0 e_2 - e_1 e_3)$$
$$\theta = \sin^{-1}(-l_3)$$

(3.20)

and since, by definition, θ is the elevation angle of the x-axis above the horizontal plane and thus lies in the range $\pm\tfrac{1}{2}\pi$

$$-\tfrac{1}{2}\pi \leqslant \theta \leqslant +\tfrac{1}{2}\pi$$

(3.21)

The inverse sine process will then yield a unique value for θ. Next

$$\cos\theta \cos\psi = l_1$$
$$\cos\psi = l_1/\cos\theta$$

(3.22)

and since $\cos\theta$ is always positive for the defined range of θ, then

$$\cos\theta \sin\psi = l_2$$

(3.23a)

yields

$$\text{sgn}[\sin\psi] = \text{sgn}[l_2] \tag{3.23b}$$

and hence

$$\psi = \cos^{-1}(l_1/\cos\theta)\cdot\text{sgn}[l_2] \tag{3.23}$$

Similarly

$$\cos\phi = n_3/\cos\theta$$

$$\phi = \cos^{-1}(n_3/\cos\theta)\cdot\text{sgn}\,[m_3] \tag{3.24}$$

To summarise the two approaches to aircraft orientation, the Euler technique is simple, widely known and widely used. Its one weakness is the singularity at zenith and nadir. The quaternion technique is more complex and less familiar but is robust and will cope with all kinds of manoeuvres.

It is interesting to compare the solution sequence in each case.

Euler:

Body rates → Euler rates → Euler angles → direction cosines
         → transformation matrix

Quaternion:

Body rates → quaternion rates → quaternions

→ direction cosines ↗ Euler angles
                    ↘ transformation matrix

3.3.7   *Axes and frames of reference*

Sets of axes are necessary as frames of reference for equations of motion, for specification of aerodynamic forces, and for definition of aircraft attitude. There are many possible axis systems, but for simulation they fall into two broad classes, 'body' axes and 'earth' or 'inertial' axes.

'Body' axes consist of an axis system fixed in the aircraft, and normally with origin O at its centre of gravity. Body axes thus move in space and rotate with the aircraft. For a specific set of body axes it is essential to know how the orientation of the axis system is defined with respect to the aircraft.

Geometric-body axes are axes defined such that the x-axis is aligned with a geometric feature, such as the fuselage reference line (FRL) or a wing datum plane (WDP). Often such axes are known just as body axes. Both kinds of reference line may be used for one model: α_{WDP} to define the aerodynamic forces but α_{FRL} for all other uses, including axis transformations. Typically α_{FRL} and α_{WDP} differ by a small constant angle, e.g.,

$$\alpha_{\text{FRL}} = \alpha_{\text{WDP}} - 2 \tag{3.25}$$

An alternative set of body axes is known as principal-body axes, defined such that the x-axis is aligned with the principal inertia axis, thus making the cross-product of inertia I_{zx} zero. This simplifies the lateral equations of motion (see equations (3.3)) but is not particularly helpful in simulation, as the location of the principal axis could change with aircraft configuration. It is, however, sometimes useful in theoretical studies.

Geometric-body axes offer sufficient advantage in giving constant moments of inertia.

Aerodynamic-body axes (sometimes also known, loosely, as stability axes in the USA or wind axes in the UK) have the x-axis aligned with the projection of the velocity vector, in a datum flight condition, on to the plane of symmetry. Aerodynamic-body axes are therefore at an angle α relative to geometric-body axes. These axes are usually employed to define the basic aerodynamic forces, especially lift and drag, as they are the 'natural' axes for such definition. Conversion of data to other axes is a straightforward matter (ESDU 1976a).

If the x-axis is always coincident with the projection of the velocity vector, then the axes are no longer body axes, because they are no longer fixed in the aircraft.

Flight-path and air-path axes also exist. Neither is a body axis system. For both, the x-axis is directed along the path followed by the origin O (the centre of gravity): either the path relative to earth or the path related to the air-speed vector. They differ only if there are winds.

'Earth' or 'normal earth' axes comprise a set of axes defined with respect to the earth, with the origin at a suitable point, such as a runway threshold, the x-axis pointing North, the y-axis East and the z-axis down. Unlike body axes, earth axes are inertial axes. They are used as reference axes for position and attitude and for the definition of winds.

Geometric and kinematic relationships for various axis systems are usefully summarised in ESDU (1976b).

3.3.8 *Incidence angles*

Unlike attitude angles, which define the orientation of an aircraft with respect to a set of earth axes, incidence angles define the direction of the airflow relative to the aircraft's body axes and are needed in the derivation of the aerodynamic forces acting on the airframe. Of the possible definitions (ESDU 1967c), the commonly used ones are the angle of attack α_t and the angle of sideslip β_s, usually written without the suffixes. They are defined as (see also Fig. 3.5)

$$\left.\begin{array}{l} \tan\alpha_t = w/u \\ \sin\beta_s = v/V \end{array}\right\} \tag{3.26}$$

Fig. 3.5. Definition of incidence angles.

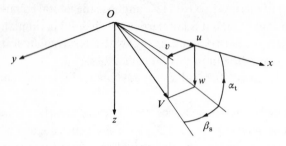

where $V^2 = u^2 + v^2 + w^2$

α_t lies in the range $\pm \pi$ and takes the sign of w,
β_s lies in the range $\pm \pi/2$ and takes the sign of v.

3.3.9   *Direction angles*

Direction angles define the direction of the flight-path velocity vector with respect to earth axes. The angles, shown in Fig. 3.6, are the climb angle γ and the track angle χ. The difference between track angle and heading is called drift and is due to the presence of winds.

If the flight-path velocity is V^k then the components of velocity in earth axes are

$$\left. \begin{aligned} u_0^k &= V^k \cos \gamma \cos \chi \\ v_0^k &= V^k \cos \gamma \sin \chi \\ w_0^k &= -V^k \sin \gamma \end{aligned} \right\} \qquad (3.27)$$

Incidence angles, attitude angles and direction angles can be related, but

Fig. 3.6. Definition of direction angles.

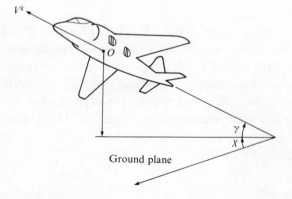

Ground plane

the expressions are complex. In the absence of winds, and in steady, straight symmetric (i.e. wings-level) flight, it can be shown that

$$\gamma = \theta - \alpha \tag{3.28}$$

but it must be emphasized that this useful and helpful relationship is only true in the restricted but common circumstances quoted.

3.4 Representation of aerodynamic data

3.4.1 *Background*

Depending on the implied form of the model, i.e. whether it is a 'total force' representation, or whether it is based on small perturbation equations, aerodynamic data may be represented in a variety of ways. For a small perturbation model, the coefficients of the equations of motion are the classical stability derivatives, and a compact table of numbers is sufficient to represent the aircraft at one flight condition. These may amount to 6 force derivatives and 15 moment derivatives: a total of 21 data items making a highly compact model.

A simulation requirement which calls for the pilot to fly the aircraft over a wide speed range, possibly in combination with a variety of manoeuvres involving a substantial variation in angle of attack, demands a complex representation of the aerodynamic forces. Forces and moments may need to be expressed as functions of one or more of the following:

- angle of attack
- control deflections
- speed/Mach number
- rotation rates
- height
- centre of gravity position
- ground proximity
- geometry (e.g., flap setting, wing sweep)

Depending on the variety of tasks to be flown, it is possible to choose an appropriate level of complexity in model representation. This is an area where the intended application of the model can have a big influence on the nature and form of the model and also on the scope and accuracy of the data.

In the early stages of the design of a new project, there will be considerable doubt about the mathematical model, and gross variations in parameters may be made to identify sensitive areas and to show where effort should be concentrated to refine both the form of the model and the data. Simulation is invaluable here to evaluate the effectiveness of the predicted performance and proposed operational use of the new design.

At the detailed design stage, a simulation can become a focus for airframe and control system designers, providing a means to evaluate alternative design strategies. This may often demand detailed and accurate modelling of the aircraft and its systems.

3.4.2   *Polynomial fits to data*

A typical plot of lift coefficient C_L versus angle of attack α for a transport aircraft is shown in Fig. 3.7. It is possible to fit the data curve with a straight line over a useful range of α values, as in Fig. 3.7. The point circled is that appropriate to the normal operating speed

$$V_{\text{ref}} = 1.3\, V_{\text{stall}}$$

so that

$$C_{L_{\text{ref}}} = (1.0/1.3^2)C_{L_{\text{max}}} = 0.59\,C_{L_{\text{max}}} \tag{3.29}$$

C_L could therefore be represented as the simplest possible expression

$$C_L = a_0 + a_1\alpha \tag{3.30}$$

This equation gives a good fit to the data in this example to nearly $0.9\,C_{L_{\text{max}}}$,

Fig. 3.7. Typical plot of lift coefficient, C_L, versus angle of attack, α, for a transport aircraft.

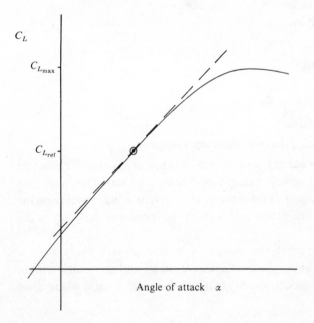

or down to a speed approaching $1.05V_{\text{stall}}$, so that if the stall region was not a relevant area for study, this representation for C_L would be quite adequate. Different flap setting will change the lift curve, but these can often be represented by simple extensions to the above formula, such as

$$C_L = (a_0 + b_0\delta_f) + (a_1 + b_1\delta_f)\alpha \tag{3.31}$$

If the changes in the basic coefficients due to flaps are not amenable to representation by such simple expressions, it may be necessary to resort to a tabular form for a_0 and a_1, e.g.,

flap	0	10	20	30 degrees
a_0	− 0.04	0.13	0.35	0.87
a_1	0.085	0.09	0.1	0.1 per degree

This approach has the benefit that once a flap change has taken place, a simple continuous expression is still available for computation. For flap angles between the defined values, a technique is required to interpolate appropriate values of the dependent coefficients a_0 and a_1. This is often referred to as 'table look-up'.

3.4.3   *The simple theory of table look-up*

Given a table of values of one variable Y as a function of a set of values of X, which plotted, look like Fig. 3.8 then the value of Y for any general value X is obtained by interpolation between adjacent coordinate pairs. Thus if

$$X_2 \leqslant X \leqslant X_3$$

then

$$Y = Y_2 + \left(\frac{Y_3 - Y_2}{X_3 - X_2}\right)\cdot(X - X_2) \tag{3.32}$$

Fig. 3.8. Schematic diagram of linear interpolation.

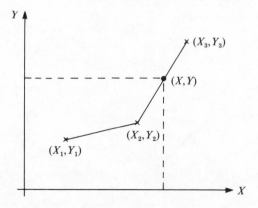

A comprehensive algorithm for this 'table look-up' process has to perform a number of steps:

(1)   decide between which pair of X values in the table the current X value lies

(2)   calculate the local slope $(Y_3 - Y_2)/(X_3 - X_2)$

(3)   apply the interpolation formula

Special action has to be taken if X is either less than the lowest X-value in the table or greater than the highest X-value. One action would be to set Y to the nearest end-point value, i.e. to Y_1 to Y_n. An alternative would be to extend the slope of the final segment, e.g. if

$$X > X_n$$

$$Y = Y_n + \left(\frac{Y_n - Y_{n-1}}{X_n - X_{n-1}} \right) \cdot (X - X_n) \tag{3.33}$$

Which of these procedures is adopted will depend on the problem and the nature of the data.

For efficient computation in a practical algorithm, it may be desirable to pre-compute all slopes, and store them as an extension to the table. This saves processing time, always important in real-time simulation. Another technique to save time is to remember the index of the lower pair of the interpolation range used last time. From one time-step to the next, the value of the independent variable X is unlikely to have changed substantially, and so it would be a good first try to use the same interval as before and thus avoid the waste of time in searching from one end of the table each time.

Function interpolation by table look-up may be, and often is, extended to functions of more than one variable. It is common to meet functions of up to four variables, e.g. lift coefficient as a function of angle of attack, elevator angle, flap angle and Mach number. Separate procedures to perform such

Fig. 3.9. Typical piece-wise linear function.

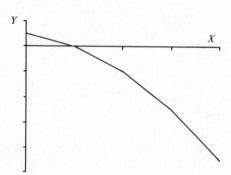

function generation calculations for 1, 2, 3 or 4 independent variables will normally be a standard feature of a simulation laboratory and its computational tools. Algorithms and implementation need to be efficient as the computing task is demanding and a large model could have many functions of up to four variables.

There are pitfalls in table look-up processes which must not be overlooked. With a relatively sparse set of data points and linear interpolation, the effective curve can look like Fig. 3.9. If the slope of each segment is plotted, it looks like Fig. 3.10. Depending on the nature of the Y-function, this can have serious consequences. If Y in Fig. 3.9 is pitching moment coefficient C_m and X is angle of attack α then Fig. 3.10 represents $C_{m\alpha}$ versus α. Now since the aircraft's longitudinal stability is strongly dependent on $C_{m\alpha}$ (or, more strictly, on the ratio of slopes $C_{m\alpha}/C_{L\alpha}$) there will be major discontinuities in a primary parameter. The significance can be reduced by having smaller intervals between X-values or by employing a higher order interpolation technique, such as cubic splines, to maintain continuity of slope. It may be better still to use a hybrid technique, with C_m as a polynomial expression:

$$C_m = a_0 + a_1\alpha + a_2\alpha^2 \tag{3.34}$$

and a_0, a_1, a_2 as $f(\delta)$, where δ may be flap angle or Mach number, for example. If δ is slowly varying (which is often true for flap angle or Mach number) the stability effects will not be significant, while the dominant effect is accurately, as well as smoothly, represented since

$$C_{m\alpha} = a_1\alpha + 2a_2\alpha. \tag{3.35}$$

One virtue of the full table look-up and interpolation technique is that it ensures that, at the set values of the independent variable or variables, the dependent variable does match the original data exactly. This may be

Fig. 3.10 Slope function derived from Fig. 3.9.

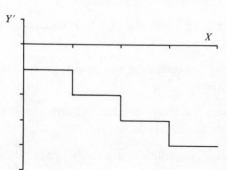

important in getting the correct aircraft performance through proper values of C_L and C_D, for example.

3.4.4   *An example of a detailed model*

A published model (Hanke & Nordwall 1971) for the Boeing 747 forms a useful example and shows the level of complexity that may be employed in practice. Some extracts are given here.

The aerodynamic lift force coefficient is represented by the general equation for a given angle of attack:

$$C_L = C_{L_{Basic}} + (\Delta C_L)_{\alpha_{WDP}=0} + \Delta\left(\frac{dC_L}{d\alpha}\right)\alpha_{WDP}$$

$$+ \frac{dC_L}{d\dot{\hat{\alpha}}}\left(\frac{\dot{\alpha}\bar{c}}{2V}\right) + \frac{dC_L}{d\hat{q}}\left(\frac{q\bar{c}}{2V}\right) + \frac{dC_L}{dn_z}n_z + K_\alpha\frac{dC_L}{d\eta}\eta$$

$$+ K_\alpha\frac{dC_L}{d\delta_{EI}}\delta_{EI} + K_\alpha\frac{dC_L}{d\delta_{Eo}}\delta_{Eo} + \Delta C_{L_{sp}}$$

$$+ \Delta C_{L_{oa}} + \Delta C_{L_{gear}} + \Delta C_{L_{ge}}$$

Each item in this complex expression is defined below:

$C_{L_{Basic}}$	basic lift coefficient for the rigid aeroplane at $\eta = 0$ (stabilizer) in free air and with the landing gear retracted
$(\Delta C_L)_{\alpha_{WDP}=0}$	change in basic lift coefficient at $\alpha_{WDP} = 0$ due to aeroelasticity (WDP = wing design plane)
$\Delta\left(\dfrac{dC_L}{d\alpha}\right)\alpha_{WDP}$	change in basic lift coefficient due to the aeroelastic effect on the aeroplane's basic lift curve slope
$\dfrac{dC_L}{d\dot{\hat{\alpha}}}\left(\dfrac{\dot{\alpha}\bar{c}}{2V}\right)$	change in basic lift coefficient due to the rate of change of angle of attack
$\dfrac{dC_L}{d\hat{q}}\left(\dfrac{q\bar{c}}{2V}\right)$	change in basic lift coefficient due to pitch rate
$\dfrac{dC_L}{dn_z}n_z$	change in basic lift coefficient due to aeroelastic inertia relief caused by normal load factor
$K_\alpha\dfrac{dC_L}{d\eta}\eta$	change in basic lift coefficient due to change in stabilizer angle from $\eta = 0$
$K_\alpha\dfrac{dC_L}{d\delta_{EI}}\delta_{EI}$	change in basic lift coefficient due to change in inboard elevator angle from $\delta_{EI} = 0$
$K_\alpha\dfrac{dC_L}{d\delta_{EO}}\delta_{EO}$	change in basic lift coefficient due to change in outboard elevator from $\delta_{EO} = 0$

Fig. 3.11. Lift coefficient, C_L, versus angle of attack, α_{WDP}, for a transport aircraft at low speed and various flap angles.

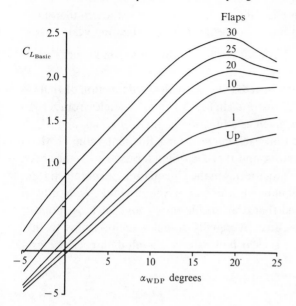

Fig. 3.12. Lift coefficient, C_L, versus angle of attack, α_{WDP}, and Mach number, M, for a transport aircraft.

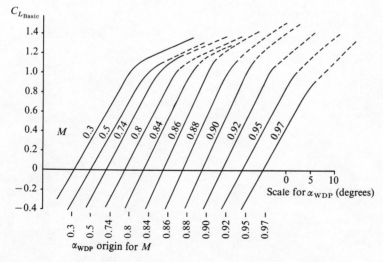

$\Delta C_{L_{sp}}$ change in basic lift coefficient due to spoiler or speed brake
            deflection
$\Delta C_{L_{oa}}$ change in basic lift coefficient due to outboard ailerons
$\Delta C_{L_{gear}}$ change in basic lift coefficient due to landing gear
$\Delta C_{L_{ge}}$ change in basic lift coefficient due to ground effect

where $C_{L_{Basic}}$ is itself, at low speeds ($M < 0.3$) a detailed function of angle of attack α and flap angle, as illustrated in Fig. 3.11 for flap angles from zero to 30°. At high speeds C_L is given as a function of angle of attack and Mach number, Fig. 3.12. Note that each curve is for a different value of Mach number, at unequal intervals, and that the origin for each curve has been displaced for clarity. The other terms in the C_L equation are obtained from further data plots and equations, which are given in Hanke & Nordwall (1971). When it is realised that drag coefficient C_D and pitching moment coefficient C_m are represented by equally complex relationships to that described above for C_L, it is clear how extensive a full data package for a modern aircraft can be.

4

Simulation of aircraft systems

4.1   Introduction

Chapter 1 drew attention to the fact that contemporary methods of flight simulation involve the modelling and reproduction of a large number of interrelated elements which represent the aircraft and the environment in which it will operate.

This chapter now examines some of the more important hardware and software aspects involved in the design of simulations of aircraft systems and, with the use of examples, attempts to provide an insight into several important areas of flight simulation.

Many factors influence these decisions, such as cost, requirements for simulated malfunctions of the equipment, complexity of the special electronics needed to provide an interface to the simulator, other hardware and displays, etc. These factors are discussed in following sections, entitled Instruments and Black Boxes.

If the decision is made to simulate equipments, this will usually involve supply by the aircraft manufacturer of visible parts such as lights, switches, control levers, panels, control box front faces, etc. The simulator manufacturer then provides the interface with the computer and designs software to model the equipment not visible to the aircrew. A brief description of the techniques and methods that may be employed to create the mathematical models and algorithms is given in subsections headed Modelling, Logic circuits, Radio aids, Engines, Control loading and Sounds.

Once the models have been established, the technology that is to be used to simulate the models must be selected. Normally, digital techniques are chosen as the most suitable technology available. The subsection headed Hardware and Software discusses the equipment and software structures for simulation.

4.2   Instruments

In modern aircraft, such as the Boeing 767, an increasing number of instrument displays are electrically driven or electronically generated and computer controlled. Such displays are normally purchased directly from manufacturers and pose only the need to interface with the simulator computer via a specially designed electronic or computer to computer link. (See later discussion of 'Black Boxes').

However, aircraft safety considerations require that in the event of electrical failure, essential flight and navigation information is still available to the pilots, via instruments sensing natural phenomena. Such phenomena are centrifugal force, gravity, earth's magnetic field, air pressures and gyroscopic effect, none of which can be adequately reproduced in a flight simulator because of conflicting sensory priorities, e.g., the sensation of forward acceleration is simulated by realigning the gravity vector by pitching up the motion platform on which the flight deck is mounted, thereby rendering impossible the use of the aircraft instrument which has a stable gyro as a pitch reference. Instruments in this category include the standby magnetic compass, using magnetic field forces, ball-in-tube inclinometers, using gravity and centrifugal force to display lateral force, standby altimeter, airspeed indicator and rate-of-climb indicators, sensing air pressure and pressure differential and a standby attitude indicator, using a stable gyroscope as a reference.

In the flight simulator these sensing devices are substituted by an ac angular positioning answering element (Synchro) or small dc servo with a dc voltage potentiometer.

Fig. 4.1. Simulation of standby magnetic compass.

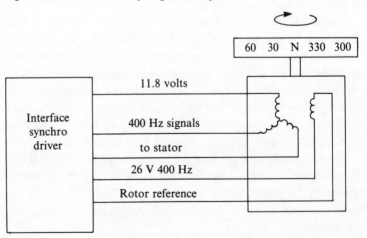

As an example of this in its simplest form, consider the standby magnetic compass. The aircraft's magnetically polarised compass card is replaced with a non-magnetic version mounted on the rotor shaft of an ac three-phase synchronous positioning device (Synchro) which responds to three related 400 Hz inputs, $V \sin \theta$, $V \sin(\theta + 120)$ and $V \sin(\theta + 240)$ (Fig. 4.1).

Consider also, an airspeed indicator having a range of 420 knots over a scale angle of less than 360°. In the simulator, the aircraft instrument's pressure capsule is replaced with dc servo assembly in which the output from an amplifier, unbalanced by a signal from the computer interface, is used to drive a pointer over the dial and position a potentiometer accordingly. The output from the potentiometer is summed at the amplifier, in opposition to the signal. Signal and answer are scaled so that when the desired airspeed is indicated, the amplifier output is 0 volts (Fig. 4.2).

It will be seen that where continuous rotation of a display is required, an angular positioning device is used due to the angular limitations of the resistive element within a potentiometer, whilst where a display excursion of less than 360° is required the latter is often used.

Whilst the above illustrates two frequently used methods of instrument drive there are several other types of instrument such as temperature indicators, which sense change in a resistive element. Means are found to drive all such instruments by synthesizing the drive voltages and currents via specially designed interface equipment.

Chapter 5 gives a more extensive list of drive techniques.

Fig. 4.2. Simulation of airspeed indicator.

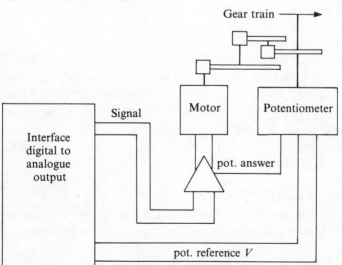

4.3   Black Boxes
'Black Box' is the term used generally to describe any of the on-board computers used in modern aircraft. A choice must be made as to whether to use the actual unmodified on-board computer (Black Box) or a software model of the same, and in order to demonstrate the criteria used in making this choice, the simulation of two of these systems is considered in turn.

4.4   Inertial navigation system (INS)

4.4.1   *Description*
The INS includes a stable platform composed of gyroscopes and accelerometers. The gyros maintain the platform's accelerometers in a level position in relation to the earth's surface, so that they accurately measure aircraft accelerations. A digital computer connected to the platform integrates the aircraft acceleration data to produce aircraft velocities and computer platform heading, resolves the velocities through the heading angle to produce North/South and East/West components of velocity which are then integrated once more. The resulting distance is added to a datum latitude and longitude entered into the computer at the start of the flight to give an accurate Present Position.

The computer also controls the initial platform alignment (levelling and orientation) and continuous corrections to the platform during flight.

The platform and computer comprise the Inertial Navigation Unit (INU). Data entry and display is made via a cockpit mounted Control/Display Unit (CDU).

4.4.2   *Simulation*
The unmodified INU cannot be used on the simulator, since the platform would not be subject to realistic aircraft motion.

The choice, therefore, is either to use the computer of the INU interactive with a software simulation of the platform running in the simulator host computer, or a software simulation of the INU. In either case, an unmodified CDU would probably be used due to considerations of authenticity and cost.

4.4.3   *Criteria*
The data supplied by the INU to cockpit displays and autopilot is not extensive, being with respect to aircraft attitude and lateral navigation only.

All input data required by a software simulation of the INU is either readily available from the simulator software programs in which it is computed (e.g., aircraft attitude and position), or can be computed using vector analysis in a close approximation to the methods used in the computer of the INU.

INS manufacturer's data to support a software simulation of the INU is readily available in the form of Pilot's Operational Notes, Maintenance Manuals, Overhaul Manuals and Technical Bulletins.

The displays of the CDU are in-line digital readouts. Data in support of these displays is easily encoded to the requirements of the CDU.

Labour costs in producing a software simulation are competitive with the cost of aircraft units, particularly after the production of an initial system.

The INU computer software is not dedicated to any particular aircraft airframe/engine configuration. Redesign of the simulated INU computer for each aircraft type is not necessary.

Core and time requirements in the simulator computer for the software simulation are low enough to be acceptable.

4.4.4   Conclusion

There is overwhelming evidence in favour of a software simulation of the INU on the grounds of data availability, ease of up-date, cost and reasonable simulator computer core and time requirements.

4.5   Flight management system

4.5.1   Description

The Flight Management Computer System (FMCS) hardware consists of two Flight Management Computers (FMC) and two CDUs.

The CDU is the interface between the pilot and the FMC and incorporates a complex character crt display.

The FMCS accepts information from other aircraft systems such as the INS, together with data from radio navigation, engine and fuel sensors. It also contains pre-flight loaded navigation and aircraft performance data. The crew enter data to perform the following functions:

navigation

guidance

performance management

Electronic Flight Instrument System (EFIS) support

data display

fault data storage for relay to the Maintenance Control Display Panel

Outputs from the system include command data to the Flight Control and Thrust Management Computers for automatic guidance, while navigation data is displayed on the Electronic Horizontal Situation Indicator (EHSI) of the EFIS. In its most advanced form this display consists of a map with an aeroplane symbol, planned flight path, and ground facilities of interest, such as navigation aids and airports. Alphanumeric information is entered and displayed on the CDU.

4.5.2   Simulation

The CDU and EHSI are cockpit-mounted units and must be authentic in appearance and operation. Complexity and cost make the manufacture of replicas uneconomical and unmodified aircraft units are used. The choice is therefore between the use of unmodified aircraft FMCs or a software simulation of the FMCs, running in the simulator computer.

4.5.3   Criteria

A software simulated FMC would give low cost repeatability for a program to identical or near identical specification.

The dual FMC installation could be simulated by an indexed, two-pass entry to the simulated FMC software program, whereas using unmodified aircraft FMCs requires that two such units be installed for a full simulation of the system, with attendant costs.

The FMC includes a microprocessor with a total memory of over 150K words to enable the storage of large amounts of data (e.g., all data for every radio navigation station and airport in North America), and a large Performance Data Base containing airframe and engine data.

The Navigation Data Base is designed to enable update via a cassette loader. Update is made on a 28 day cycle, using cassettes provided by a specialist supplier.

Development of the FMC for a highly competitive market has taken place over a number of years and represents a considerable investment by the manufacturer. Details of the unit are therefore highly proprietary. Obtaining data in support of a system simulation therefore presents serious difficulties.

Development of the FMC is rapid and continuous. To keep abreast of developments, a parallel development program of the simulated system would be necessary.

The FMC is supported by purpose designed test equipment supplied by the manufacturer.

4.5.4 Conclusion

There is overwhelming evidence against a software simulation of the FMC on the grounds of:

> Cost of initial design/development including that for reconfiguring the data from the loader format to that required by an 'own' design, navigation data base

> on-going update development cost and the difficulty of obtaining data memory/computer time requirements

4.6 Modelling

In the context of simulation, modelling means the design of mathematical equations which produce numerical representations of the real world under equivalent input conditions. Because space does not allow specific discussion of the many aircraft systems the three main categories of modelling are discussed here, supported by examples in the following subsections entitled Logic Circuits, Radio Aids and Engines.

4.6.1 Component models

These are produced by consideration of components such as:

bus-bars, circuit breakers, switches etc. – electrics

pumps, valves, tanks, reservoirs etc.   – hydraulics/fuel

compressors, valves etc.   – cabin conditioning

Schematic diagrams, usually partly available from the aircraft manufacturer, are drawn (as in Fig. 4.3, illustrating logic circuits in the following section).

Each component, plus flows and pressures etc. can be individually represented by established techniques/equations based upon physical laws. Thus from assembly of individual equations a model based upon the schematic/diagram of the system can be produced.

4.6.2 Input/output models

These are produced for complex equipment/hardware such as electronic Black Boxes, when the required depth of simulation can be achieved by treating these as components while largely ignoring the detailed inner working.

The design data, often in the form of graphs (such as that illustrated in Fig. 4.7 of the Engines Section) is often supplied by the manufacturer from tests done to support or certificate aircraft design. Alternatively the manufacturer supplies raw test results/curves which are analysed/approximated by the Design Engineer to produce data curves suitable for entry into a computer. Sometimes it is necessary for the design engineer empirically to design a model which has feed-back control paths. (See Fig. 4.8 of the Engines Section.) Often input/output models are mixed with component models in the overall design (e.g., in the Engine System).

4.6.3   *Mathematical models*

These are often based upon classical mechanics (e.g., equations of motion), spherical trigonometry (see later section on Radio Aids) or mathematical papers resulting from studies undertaken at universities, government agencies e.g., atmospheric turbulence has been the subject of such mathematical treatment involving the theory of non-Gaussian distribution. Depending on complexity, the design engineer expresses the model first as a schematic design which visually shows the flow path of equation outputs which form inputs to succeeding equations. The flow schematic is particularly worthwhile when there are multi- and feedback paths as with the flight equations. From the flow schematic or basic equation data, the design engineer produces an ordered sequence of equations suitable for programming.

4.6.4   *Summary*

In all types of modelling the designer initially constructs a component diagram or schematic flow diagram in a form traditional to the system. The diagram can then be studied for completeness, depth of simulation etc. and discussed with Lead Engineers. From the diagram an ordered list of equations can be produced together with a list of system input/output variables which can be checked against the corresponding variables produced and required by other systems.

4.7   Logic circuits

Whilst simple circuit logic can be modelled by if/then/else statements, complex logic is often expressed mathematically in the form of Boolean equations (See Fig. 4.3.).

The normal method of analysis would involve generating a Boolean equation for each element of the circuit, starting at the input and proceeding stage by stage to the output. The equations so generated would, if translated directly into software code, serve as a working model. However, certain other factors need to be taken into account.

Ways of reducing program execution time and memory storage requirements should be considered, e.g., although the logical expression $(\bar{A} + \bar{B}) \cdot \bar{C} \cdot \bar{D}$ is equivalent to $\overline{(A \cdot B) + C + D}$ the second expression is preferable since it involves four logic operations (one AND, two ORs and one inversion), as opposed to seven (two ANDs, one OR and four inversions).

It is also possible to save execution time by introducing logical decisions, i.e., branching, into the program. Excessive use of branching should be avoided, however, as the program tends to become disjointed. If the logical term E is equated to the above expressions it can be seen that if D is true, E must be false. It might be useful, therefore, to test the state of D and, if true, simply set E false. However, no time is saved should D be false since, E must equate to $\overline{(A \cdot B) + C}$ and four logical operations have to be performed (one AND, one OR, one inversion and one branch decision). This illustrates that a logical decision can reduce execution time under restricted conditions without reducing the most important parameter – the worst case execution time. In the above example the usefulness of introducing the decision would depend upon the nature of term D. Clearly, if D is normally false little would be gained.

4.8   Radio aids

4.8.1   *Ground station*
The basic data provided by navigation ground stations are distance and/or bearing (Table 4.1).

Fig. 4.3. Representation of logic circuits.

Note: In the modelling of above circuit, the 100 A fuse and the transformer rectifier units (TR) are assumed to be permanently fully functional.
Tie Point = (ac Bus 1.C810 + ac Bus 2.C809). dc Bus 1 ON = Tie Point·C762, dc Bus 2ON = Tie Point·C763, ESS dc Bus ON = [Tie Point·C53 + (ESS ac Bus·C811]·C764.

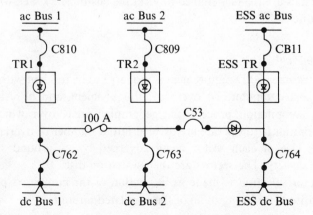

Table 4.1. *Examples of navigation ground station types*

Ground station	Abbreviations	Basic data provided
Non-directional beacon	NDB	Bearing (station w.r.t. aircraft)
VHF omni-directional range	VOR	Bearing (aircraft w.r.t. station)
Distance measuring equipment	DME	Distance (aircraft to station)

The data required to simulate navigation ground stations and a number of other radio aids to navigation are primarily:

> type
>
> frequency
>
> geographical position (latitude and longitude)
>
> height above sea level
>
> magnetic variation at ground station site
>
> identification call sign
>
> transmitter power (high, medium, low)

Ground station data (GSD) source

Data for all types of ground stations is provided by specialist companies who maintain the current status of the data by continuous update procedures and supply it to the simulator manufacturer on magnetic tape. The simulator manfacturer, using utility software programs, transcribes the data into a suitable format and produces master listings of GSD (world wide). From the master listing, the customer chooses his ground station requirements which the simulator manufacturer transfers into dedicated files. These files are then assembled to form the customer's GSD, which is stored on disc.

GSD management

The total customer GSD requirement is too large to be resident within the simulator host computer. To overcome this problem, and by using an automated compilation process, the geographic areas over which the simulated aircraft is to operate are divided into a number of overlapping sectors, the size of each sector being governed by its ground station population density. The sectors are then stored on disc.

GSD sector selection is made as a function of the computed present position (latitude and longitude) of the simulated aircraft.

Aircraft present position

Present position (latitude and longitude) of the simulated aircraft are computed by resolving simulated aircraft body axis velocities through simulated aircraft attitude angles to produce ground axis velocity.

Ground axis velocity is resolved through simulated aircraft true heading to produce North/South and East/West velocities which are then integrated with respect to time and scaled to produce nautical miles of distance travelled from the starting point. By definition, one nautical mile equals one minute of arc in latitude while one nautical mile equals 1/[cosine (latitude)] minutes of arc in longitude. These angles added to (for northerly and easterly movement) or subtracted from (for southerly or westerly movement) the starting position (latitude and longitude) of the simulated aircraft, provide a continuously updated Present Position.

Distance calculations

Since the simulation of Radio Navigational Aids is concerned only with comparatively short distances (up to 300 nautical miles) it is possible to

Fig. 4.4. Computation of distance and bearing. S = Ground Station site, A = Aircraft Present Position, λ_S = Site latitude (radians), λ_A = Aircraft latitude (radians), $\delta\lambda$ = (Aircraft latitude – Ground Station latitude) (radians), $\delta\mu$ = (Aircraft longitude – Ground Station longitude) (radians), d = Great Circle distance, S-to-A (radians), B_1 = Bearing (A w.r.t. S), B_2 = Bearing (S w.r.t. A).

Note: B_2 cannot be assumed to be the reciprocal of B_1 due to convergence of the meridians of longitude, towards the True North Pole.

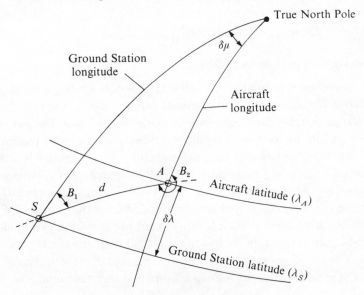

consider the earth to be a sphere and that the distance between the tuned ground station and a point on the earth's surface immediately below the simulated aircraft, will be measured along a segment of a Great Circle (see Fig. 4.4).

Spherical geometry is used to calculate the distance d, the basic equations being:

$$\cos d = \sin\lambda_S \sin\lambda_A + \cos\lambda_S \cos\lambda_A \cos\delta\mu$$

By expansion and substitution and by ignoring third and higher order terms occurring during these processes, because they are small, an expression for distance d is obtained as follows:

$$d = \sqrt{(\delta\lambda^2 + \cos\lambda_S \cos\lambda_A \, \delta\mu^2)} \text{ radians}$$

Radians are then converted to minutes of arc each of which by definition is equal to one nautical mile.

Bearing calculations.

Bearing B_1 and B_2 are obtained using spherical trigonometry.

Both the sine and cosine of B are calculated and entered into an ARC B subroutine to obtain angle B

$$\sin B = \frac{\cos\lambda_A \sin\delta\mu}{\sin d} \text{ radians}$$

$$\cos B = \frac{\sin\lambda_A - (\sin\lambda_S \cdot \cos d)}{\cos\lambda_S \cdot \sin d} \text{ radians}$$

4.8.2 *Examples of ground station simulation:*

Non-directional beacon (NDB)

The receiver is simulated in a software program. Coded, selected ground station frequency data is transmitted from the cockpit mounted control unit via suitable data links, to the simulator computer. Frequency data, decoded in the software program is used in a search by frequency, of the GSD for NDBs stored in the simulator computer. Distance d is calculated for each frequency matched NDB and compared with its transmission power (range) data, extracted from its GSD. If more than one NDB is in range, the distance d is used to determine and select the nearest NDB as the tuned station, after which it is used to control the volume of the simulated audio transmissions of the Identification Call Sign and to detect entry into the Cone of Confusion, an area centred on, and in close proximity to, the NDB in which received bearing information becomes unreliable.

Sine and cosine of bearing B_2 are calculated for the NDB since, in the aircraft system, bearing is determined by the loop aerial of the receiver which is resident in the aircraft.

Bearing B_2, calculated in the ARC B subroutine, is used to drive the Bearing Pointers of the Radio Magnetic Indicators, located on the cockpit instrument panels.

Simulated audio volume is also controlled as a function of bearing when the loop aerial is used as a manually operated Direction Finder.

VHF Omni-Directional Range (VOR)

Simulation of the receiver and selection of the tuned station from those available in GSD are similar to the methods described for the NDB. Received signals from the simulated transmitter are in this case subject to line-of-sight limitations imposed by the earth's curvature. These effects are simulated using the equation:

$$R = 1.22(\sqrt{H} + \sqrt{h}) \text{ nautical miles}$$

Where:

R is the ground range, transmitter to receiver, at which the receiver falls under the 'shadow' of the earth's curvature (nautical miles)

H is the height of the transmitter above sea level (obtained from GSD) (feet)

h is the height of the receiver (aircraft height above sea level) (feet)

Distance d is compared with R and if greater, the simulated signals from the transmitter are invalidated.

Simulated fading of the Identification Call Sign signal volume as a function of distance d is not necessary as the aircraft receiver includes Automatic Gain Control (AGC).

Referring to Fig. 4.4, sine and cosine of bearing B_1 are calculated for the VOR, since the transmissions are in the form of a variable phase signal and a reference phase signal, the relationship of which is such that when decoded by the receiver, they are indicative of the magnetic bearing of the aircraft with respect to the transmitter.

The bearing is displayed at the Radio Magnetic Indicators on the cockpit instrument panels.

The net effect of the transmissions is to produce identifiable spokes or 'radials' emanating from the transmitter at one degree intervals. The receiver enables a choice of radial on which to approach or depart from the VOR and by comparing computed bearing B_1 with the chosen radial, deviation from the radial is detected. After suitable formatting and scaling the deviation signal is used to drive indicators at the instrument panels and to provide steering information to the Automatic Flight Control System (AFCS).

Distance measuring equipment (DME)

Selection of the tuned ground station from GSD is by frequency matching as for the NDB.

In the aircraft system an airborne transmitter/receiver interrogates the DME ground station with a transmitted pulse and receives, after a short finite interval, a reply pulse. The time interval between transmitted and received pulses is a measure of the slant range, aircraft to ground station.

Referring to Fig. 4.5, slant range is calculated as:

$$\sqrt{[(\text{distance})^2 + (h - H)^2]}$$

Range is 200 nautical miles maximum and is limited by line-of-sight considerations as described for the VOR.

Display is on a digital readout or by rotary counters at the cockpit instrument panels.

Other navigational aids

All other navigational aids, such as Instrument Landing Systems (ILS), Landing Markers (LMKs), Tactical Aid to Navigation (TACAN), and Simulated Ground Control Approach Systems (GCA), are based on manipulation of the parameters distance and bearing.

4.9   Engine design

In simulating performance of those parts of an aircraft which have mechanical components, the designer is usually faced with alternative high or low sophistication approaches to simulator model design. This is particularly true of engine simulation.

Fig. 4.5. Distance measuring equipment calculations. H = Height of Ground Station above sea level, h = Height of aircraft above sea level, d = Ground distance (Aircraft to Ground Station).

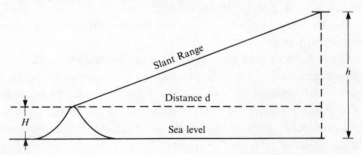

A basic gas turbine engine consists of an inlet, compressors, combustion chamber, turbines and exhaust nozzle (see Fig. 4.6).

The designer is faced with a requirement to simulate the engine throughout its operating envelope such that it will functionally interface and provide simulated failure information to all related systems.

To achieve this object there are two basic methods, and most simulations use a mixture of both methods. The variations are brought about by the constraints of available data, cost and the level of fidelity required.

4.9.1   *Total output simulation method*

In this approach the various required outputs, shaft speeds, thrust, etc., are defined as complex mathematical functions of the input conditions and are used in an open loop. The method requires a considerable volume of data plus mathematical expertise of a high order. It is difficult to incorporate failure cases for training, it is inflexible, and it is expensive in computation time. Consequently, total output simulation is normally restricted to engineering/research applications where training and real-time simulation is not a requirement.

4.9.2   *Component simulation method*

This method involves the modelling of each major component of the engine, e.g., compressor, combustor, turbine and nozzle, using theoretical thermodynamics, together with components of the ancillary systems, such as the starter and the oil system. All the component simulations are then coupled together and use the simulated fuel control unit to close the

Fig. 4.6. Schematic of gas turbine engine.

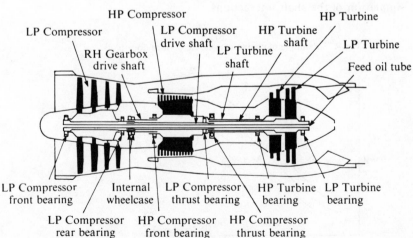

computation loop in order to provide a close analogue of the real engine. The data is derived and supplied by the engine manufacturer from engine and component tests, and parts of it are often correlated with results from the manufacturer's engineering simulation.

As for total output simulation, this method requires considerable mathematical and thermodynamic expertise of a high order. It is very expensive in computation time.

4.9.3   *Typical method*

Most simulations make use of the manufacturer's normalised steady-state performance which is a series of parameter relationships representing steady-state performance for specific input conditions, i.e., fuel flow (W_f), would be presented as a carpet plot against engine core speed (N_H), for a range of mach numbers (M) (Fig. 4.7).

Note that such graphs do not identify specific engine components but are representations of the overall performance of those components, i.e., like the total output method. The terms δ and $\sqrt{\theta}$ are normalising terms dependent upon airspeed, altitude and atmospheric conditions (pressure and temperature).

This approach allows the final stable performance to be defined as the output from a series of look-up tables. What now has to be catered for is the transient performance of the engine.

To achieve realistic transient simulation, a closed loop is formed using a model of the fuel control unit (as in the component simulation method). This is illustrated in Fig. 4.8.

For a typical engine of two-spool design each shaft is treated as shown in Fig. 4.8 and partial derivative terms are added to improve the dynamic simulation of the shaft interactions.

Fig. 4.7. Example of engine parametric performance plots.

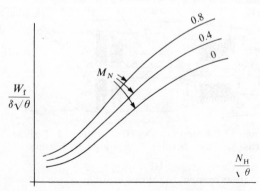

The perceived performance, as seen on the cockpit mounted instruments, using this method can be sufficiently close to a real nominal engine so that the differences can be attributed to engine-to-engine differences in real life.

However, if a high degree of realism is not required then the model can be simplified by removing the LP shaft loop, making it a fixed relationship to the HP shaft speed. This still gives correct steady-state performance but the transient performance is now only a general representation.

A further reduction can be made in both cost and complexity by reduction of the manufacturer's data, removing or simplifying temperature, altitude and speed effects, and introducing straight line approximations where possible. This approach will give a degree of realism that is acceptable for procedure training where the emphasis is not on the absolute performance but on the procedure to be carried out in order to achieve a required result, i.e., in the training of actions to be carried out that relate to the Emergency Procedures Checklist carried in all aircraft.

4.10 Control loading

The Control Loading System exists to produce feel forces, on the simulator's flying controls, which accurately reflect those felt by the pilot in actual flight conditions. In order to achieve this it is necessary to produce the correct feel force gradients throughout the entire control range of the subject aircraft. In actual flight conditions these forces are produced by aerodynamic loads on the aircraft surfaces or by artificial feel units of various types.

The pilot's controls can be divided into two categories, primary and secondary. Primary controls include the column and wheel (or stick) and the rudder pedals, and give rise to elevator, aileron, spoiler and rudder surface movements. Secondary controls encompass items such as thrust

Fig. 4.8. Engine transient response simulation. W_F: Schedule fuel flow, I_{shaft}: Shaft inertia, W_{FS}: Steady-state fuel flow, N_{shaft}: Shaft speed, τ: Torque, $TT2$: Total air temperature.

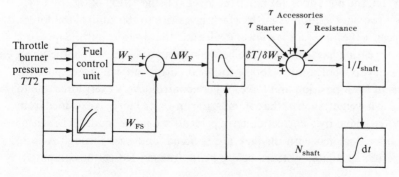

levers, nosewheel steering tiller, flap levers and speedbrake levers. Simulation of these controls may be achieved by passive or active methods, the latter being employed to simulate more realistic feel for primary systems and in a number of the secondary applications.

Passive systems utilize mechanical spring units, viscous damper mechanisms or electromechanical brake systems. These systems have in the main, been replaced by active systems using electrohydraulic techniques controlled by analogue or digital computers. The more flexible digital systems have, in many applications, superseded the analogue systems which suffered from modelling limitations and the need for special system hardware.

Active systems utilize low frictional devices, termed hydrostatic force jacks, to drive the control levers to which they are linked. Movement of the hydraulic element is controlled by a servo valve activated by a drive signal controlled by the system model. The forces applied by the pilot are measured by a load cell positioned between the lever and its associated hydraulic element Control position is usually measured by a potentiometer or optical encoder. In certain applications a velocity transducer may be incorporated. In order to limit distortions and resonances, the mechanical components are located in a rigid frame.

Before proceeding to modelling it is worth considering the effect of force, applied by the pilot, on a lever linked to a hydrostatic force jack. This force is sensed by the load cell, which is a highly accurate and linear device, the output of which forms the force term which is of overriding importance. It is used to derive a control acceleration term (by dividing by mass) which is integrated once with respect to time to provide control velocity, and further integrated with respect to time to give control position. This desired position is compared with the actual control position signal measured by the potentiometer and the error is used to energise the servo valve which produces a hydraulic flow to reposition the force jack piston which in turn moves the control lever.

In fact, the pilot does not move the lever, it is the action of the force jack which provides a 'mobile earth' which gives rise to the control feel forces.

Because the force jack is a linear device and the aircraft control inputs are rotary, gearing has to be provided between the two elements. It is, therefore, necessary to incorporate a rotary-to-linear correction which is required in terms of both position and force. Where control levers exert force due to their own weight, such as the columns, the unwanted force is excluded from the force signal by a correction term prior to use in the model. The sensor outputs associated with the jack are buffered, and pass through A-to-D converters in digital systems, before being transmitted to the computer.

To achieve the correct control feel it is necessary to produce a mathematical model of the system. The model is formulated from the aircraft manufacturer's control data, which usually includes wiring diagrams, mechanical system gearing, graphical representation of forces and control surface deflections. Also supplied, are surface hinge moment data to enable the simulation of boost-off conditions and details of malfunctions which are derived from failure analyses of the aircraft control channels. The data is analysed to enable it to be adapted into a form suitable for use in the chosen method of simulation. The major items requiring simulation are shown in Fig. 4.9.

The models are derived from analysis of the aircraft system data and the described features. Many of the functions, such as position limits and inertia have numerical values, as do control surface gearings and auto-pilot drive rates and authorities. All these values are included in the model. Probably the most important data is that given for the control centering spring which usually represents the major element felt by the pilot. In most cases the stiffness of the control run is such that the centering spring, which is usually remote from the pilot, is a well defined and identifiable feature of each aircraft type. Sometimes, the spring gradient is designed to be a function of airspeed.

Generally, second-order models are used to represent both front and aft ends of the control runs. These models are linked by a complementary cable model for one control. A good representation of the aircraft surface actuator is achieved by the use of a first-order model. Fig. 4.10 depicts

Fig. 4.9. Potential component contributions to control loading.

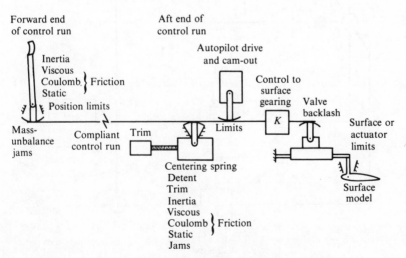

a typical model for one control. The pilot applied force is derived from the load cell and the autopilot drive is synthesized in the host computer. Where manual reversions occur, the model must provide the boost-on/off facility in order to simulate the different forces experienced.

The method of modelling is of prime importance and therefore, the selection of a suitable control loading computer must be given careful consideration. Assuming that a digital system is to be employed, it is necessary to select a computer which is sufficiently fast to cope with the equations that must be modelled. It has been confirmed by analysis of aircraft control systems, that frequencies up to 50 Hz may occur due to discontinuities or when control stops are encountered. In order to avoid

Fig. 4.10. Control loading modelling technique.

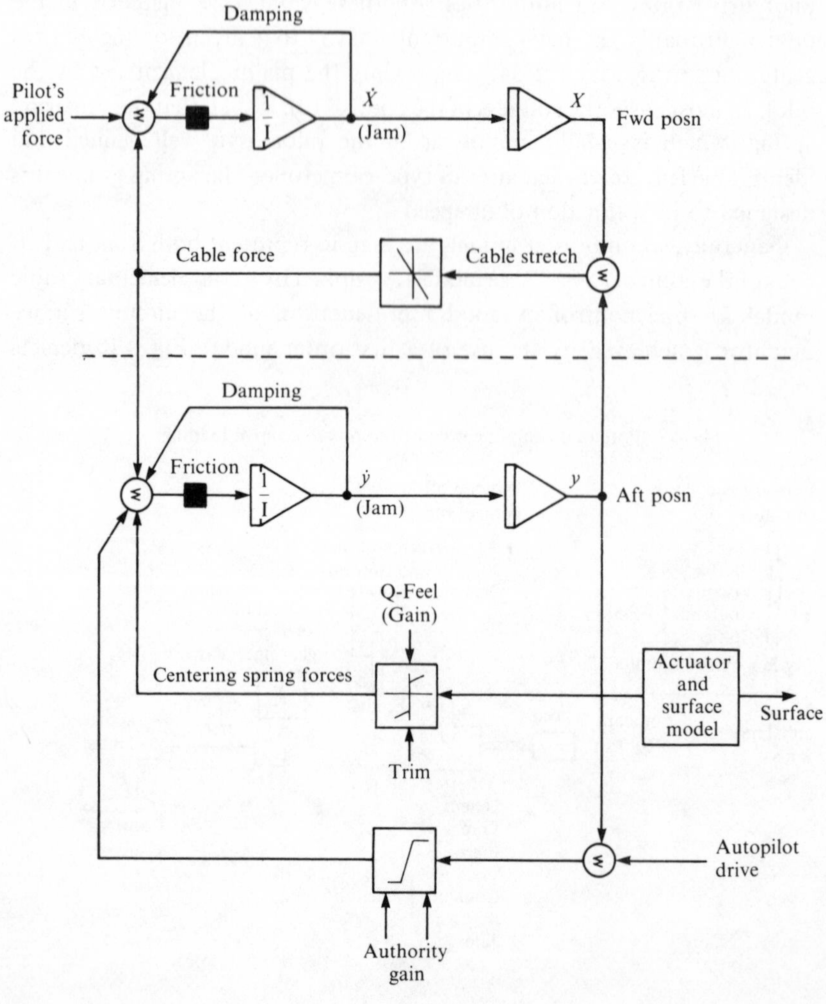

aliasing problems, a rate at least 10 times that of the highest frequencies encountered in the aircraft system should be used.

In general, it is considered that 500 Hz is the minimum iteration rate required so that the effect of simulating the control feel is integrated with the pilot's senses. This ensures that there are no discontinuities or apparent stepping of any parameters that can be felt by the pilot. The use of 500 Hz allows for accurate synthesis of the dynamic situation, such that a step input to the control column produces the relevant fast control and surface response that would be apparent in the aircraft.

It should be remembered that a control loading computer is under the influence and control of the main simulator (host) computer. The host computer will normally have a maximum iteration rate between 20 and 30 Hz; this being the rate considered most appropriate for man-in-the-loop situations. Surface data from the control loading computer is therefore supplied to the host at the maximum iteration rate of the host computer. Furthermore, the simulated aircraft, via its equations of motion, is exercised at the maximum iteration rate of the host to give instrument and visual outputs. That the control loading and host computers have different iteration rates does not give rise to any serious asynchronous problems. This is because, in general, 20 Hz is beyond human detection capabilities. Components such as autopilot drives, trim changes and atmospheric condition changes are all slower than 20 Hz.

Also, of considerable importance, are the safety aspects. This is particularly true where high speed hydromechanical devices are used. In order to protect pilots and maintenance crews, various facilities are incorporated to prevent unscheduled movements of the controls.

In addition, it is necessary to provide suitable fault diagnostic facilities and a daily readiness routine.

During the development stages of the simulation the interface between the systems must be made to function correctly and match the aircraft manufacturer's data. It is also necessary to prove that, throughout the simulated flight envelope, the quasi-static control forces plotted against positions, the dynamic response of the controls (control displacement against time), and the surface displacement due to control displacement, meet the design requirements. These requirements are effectively checked using plots and hardcopy. However, the final assessment of the system may be a subjective one made by pilots familiar with the aircraft type.

4.11   Sounds

In advanced simulators all sounds audible in the cockpit are reproduced including those which would be generated by aircraft systems (e.g., engines, air conditioning, weapons, etc.) and those which are dependent upon the simulated environment, (e.g., aerodynamic and weather sounds).

In order that the sounds may be selected and synchronised with aircraft controls and conditions, it is necessary to produce the sounds synthetically rather than to use tape recordings. The design requires a high-quality calibrated recording of the chosen sound sources. The recording is then analysed for frequencies and power spectral densities of individual components of the sounds which have been identified as relevant.

As most sounds are common to many aircraft, manufacturers have designed standard electronics which require only calibration components to provide the required range of sounds. Variation of sound in sympathy with the aircraft conditions is then controlled by inputs to the electronics derived from software which interrogates conditions, such as engine RPM, to produce corresponding variables representing frequency and amplitude.

For example, the sound of a high bypass ratio gas turbine engine may be broken down into components due to: air intake, fan buzz, compressor whines, efflux roar, etc.

The software interrogates the engine shaft speed(s) and EPR (engine pressure ratio) and computes functions of these to control amplitudes (volume levels) and frequencies (pitch) of the components via D-to-A converters. The components are summed and amplified prior to loud-speaker drive (Fig. 4.11).

The control laws are arranged to give a wide dynamic range from engine start through to full power. As the engine conditions change, so does the whole synthesis of engine sounds.

For sounds which contain detectable frequencies, analogue outputs are used to drive voltage controlled oscillators whose ranges of frequencies are selected by resistor and capacitor components on printed circuit boards.

For sounds with random frequency content, noise sources are used

Fig. 4.11. Schematic of engine sound system.

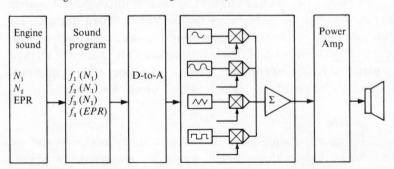

together with suitable filtering to achieve the required frequency characteristics. Several types of filters are used including low pass, band pass, high
pass, notch and dynamic filters. Each can be adjusted for the requisite
characteristics by the selection of resistors or capacitors.

Transient sounds (such as nose gear downlock) are produced by
modulating an appropriately filtered noise source with an envelope created
by a 'pulse-shaper' which is fired from the computer to initiate the transient
sound.

All individual sounds are then amplitude controlled by modulators
driven from software controlled analogue outputs. These outputs are
filtered to remove any staircase effects upon sound levels.

All sounds are summed prior to power amplification, and it is at this stage
that mixing is introduced to achieve the required directionality of the
simulated sounds.

Loudspeakers are arranged around, above and below the simulated
flight deck to achieve a high-fidelity polyphonic environment. Typically six
directions, with respect to the pilots, may be required to achieve sufficient
directionality of the sounds without incurring the unwanted effect of a
'point source.'

4.11.1 *Towards the future – digital synthesis*

With the advent of digital signal processing chip sets and bit slice
architecture, the viability of a full frequency, totally digital sound system is
assured.

The benefits from such a digital system will be compatibility of hardware,
improvement in fidelity of spectral content and amplitudes, together with
ease of expansion to add extra sounds.

4.12 Hardware and software

While simulation can be achieved using either analogue or
digital techniques, it is the latter which is usually chosen by the industry as
the most suitable and which is described in this chapter.

To maintain true, accurate and faithful simulation it is necessary for the
computer to:

perform a vast amount of computation requiring Computer Time

access a vast amount of data stored in Memory

communicate a vast amount of data between peripheral devices
and other computers and the cockpit hardware.

A typical computer configuration, which has a main simulation computer (host computer) and satellite computers is shown in Fig. 4.12. The

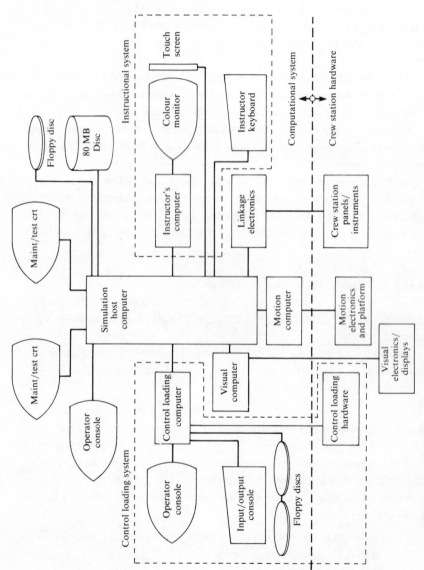

Fig. 4.12. Typical computer configuration.

host computer is the controlling computer for the satellite computers which in turn provide motion, visual, instructor's facilities and control loading. In addition, the main computer drives all of the utility general purpose peripherals as well as the dedicated flight deck peripheral. Fig. 4.13 depicts a typical sequence of events with respect to an individual simulation problem. A force applied to the control column is processed in the control loading computer, passed to the main computer, where, after considerable computation, it is converted to the necessary signals to drive the appropriate flight deck instruments, and provide the motion and vision cues.

The software structure is very much determined by the computer architecture, but generally, the computer hardware supplier provides a commercial operating system/monitor which is used to respond to hardware interrupts and pass control to the software 'tasks'. Since aircraft simulation is a very large program and demands accurate synchronization with actual real-time there is an 'executive' system within each software task associated with simulation. Each task is structured to have sections of code that have individual entry/exit points so that sampling of each section can be independently determined by the executive which controls the frequency of execution.

In the process of detail design, simulation software is further subdivided into system components, some of which model recognisable equivalent

Fig. 4.13. Block schematic of simulator interconnections.

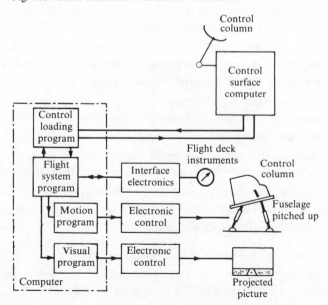

systems in the aircraft, e.g., engines, while others compute the software drives to motion, sound and visual hardware.

These systems are further divided, where possible, into sub-systems such that each defines a single real world problem. Once the systems/problems have been sub-divided, these divisions are converted into units of computer code (software modules).

Note that there are occasions where the problem is large and several software modules, sometimes running at different execution rates, are used in the solution.

4.12.1   *Time*

Simulation problems must be dynamically modelled, to ensure that all cues, displays and interactions between the machine and the operator occur in real-time. Experience and medical evidence show that, when a pilot is flying an aircraft, the threshold at which human sensory organs detect a non-continuous effect varies depending on rate of change, but typically, when a rapidly changing effect is refreshed below a rate of 20 Hz. An exception to this is control feel, in which a pilot can detect a non-continuous effect with a refresh rate of less than approximately 500 Hz.

Thus the real-time simulation systems consisting of groups of software modules are designed to be iterated by the computer at a rate which is selected to suit the simulation task. The rate that any system or module is iterated is software controlled.

The execution rate for each module must be chosen on the basis of need because computer time is limited and costly. The following computations must all be given special entry rate consideration:

> those modules which compute terms which change directly as a result of pilot or autopilot control movements and provide outputs which are part of a closed control loop
>
> those modules which compute terms in a feedback loop which, due to rapid response, are subject to instability if not computed fast enough e.g. engines
>
> those modules which compute rapidly changing aircraft parameters where time delay would be noticed in the instrumentation.

Conversely, slowly changing parameters – such as outside air temperature (dependent on altitude) – and those not in a fast response feedback loop, may be computed at slower rate.

4.12.2   *Simulator executive and rate sequencer*

Modern training simulators are designed to have a maximum iteration time interval between 33.3 ms and 50 ms depending on manufacture. This means that the fastest computation can take place at a frequency of 30 Hz or 20 Hz, respectively. (Note that this does not apply to specialised computation required for control feel and digital sound simulation which requires much higher iteration rates.) It is not necessary to compute every parameter at this frequency so it is usual to provide the facility for many of the software modules to run at sub-multiple rate, e.g., 30, 15, 7.5 and 3.75 Hz. Usually some software needs to run synchronised with special hardware such as that found in aircraft equipment and a synchronous rate entry whose time of entry is controlled to an exact time interval is provided for this purpose.

Software entry sequence is carefully scheduled in an order to ensure a minimum latency in the calculations, particularly those that provide responses to the pilot resulting from control movement (e.g., visual).

The simulator executive software schedules the software running in each RATE entry as in Fig. 4.14. After all computation scheduled for the iteration is complete it returns control to the computer operating system which can then schedule background programs. However, these are suspended when the next interrupt occurs signifying the start of the next simulation iteration.

The Executive also looks after other activities such as providing each program with the actual TIME since it was last entered.

4.12.3 *Memory*

The main computer often consists of several processing units to complete the necessary work that needs to be done in any given cycle and it is necessary that they communicate with each other efficiently. When the

Fig. 4.14. Simplified rate structure chart.

Iteration	1	2	3	4	5	6	7	8
Max rate	1	1	1	1	1	1	1	1
Half rate 1	1	–	1	–	1	–	1	–
Half rate 2	–	1	–	1	–	1	–	1
Quarter rate 1	1	–	–	–	1	–	–	–
2	–	1	–	–	–	1	–	–
3	–	–	1	–	–	–	1	–
4	–	–	–	1	–	–	–	1
etc.								

computers are of a similar type, usually this means produced by the same manufacturer, a system of communication using shared memory is the most efficient method. Shared computer memory is wired electronically to allow several computers to be connected directly into it, i.e. multi-ported. There is an impact on the speed of the computers when accessing the shared memory area. This is because when two or more computers attempt to access the same area of memory at the same time, only one will be successful, the others wait their turn on a priority basis. However, this method is favoured by most designers since it is simplest to control from a software viewpoint.

The main computer and the satellite computers may be of a different type and/or manufacture. Therefore, shared memory may not be available and communication is performed using a computer to computer link which is software controlled. Information is buffered in memory at both ends and sent to the other machine by a high speed data transfer every computer iteration. The actual transfer is automatic and runs in parallel with other computation once it has been started. A system of 'hand shaking' ensures that the transfer has been completed and indicates to the scheduler that the channel is free for new traffic.

Communication between modules of the same system is often accomplished by using local system common memory. However, if one system module needs to communicate with another, then it is usual to communicate via a global area of memory which can be accessed by all systems.

4.12.4   *Global communication areas and input/output*

Data is available to modules and to the hardware via an area of memory assigned for this purpose, e.g., the engines simulation system computes the value of the Exhaust Gas Temperature (EGT) and places it in a memory location in the global data area. Subsequently, this data is transferred out to the fuselage by a separate task dedicated to the input/output activity.

The simulation programs compute variables which are required by other aircraft systems. Typically, 'engine thrust' is computed by Engines and stored in the global data area. Subsequently, when the Flight System Program is executed it can refer to the global area to pick up the current value of 'thrust' and use it to compare with computed 'drag' to determine aircraft accelerations (Fig. 4.15).

Variables residing in the global/common areas are allocated storage by including them in a DATA BASE. The data base facility software processes the total requirement off-line and allocates the global memory area – sometimes called DATAPOOL. When an aircraft system is 'compiled' this creates address links to the DATAPOOL information.

This area is also used to form block buffers of information which are transferred to the fuselage via linkage outputs, to magnetic disc or tape, to other computers, or to special equipment, e.g., via serial data outputs.

4.12.5 *Real-time subroutines*

Since many of the aircraft simulation systems are performing similar tasks, memory is saved by the use of common routines known as subroutines which can be used on demand by the system programs. Subroutine libraries are made available on all simulators for many standard mathematical computations plus a number which are unique to simulation. Standard mathematical subroutines include sine, squareroot, integration, etc., whilst unique simulation subroutines perform standard conversions, data limiting and searching, function generation, etc. Separate

Fig. 4.15. Typical computer memory arrangement.

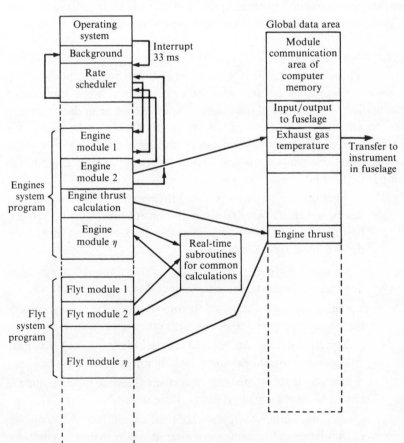

subroutine libraries also normally exist for peripheral control and off-line utility usage.

Because during each iteration subroutines are 'called' many times, special care is taken during design to limit their computer time requirements.

4.12.6   *Real-time control of computing peripherals*

Some simulator users require facilities which record large amounts of data during the simulator operation. This data is then used later to analyse pilot performances and to demonstrate any useful research or training features. To achieve the large data storage, information is written to mass storage peripherals such as magnetic disc or tape.

The simulation programs also require data which is stored on disc to be made available in real-time. Typically weather, ground station information and Instructor VDU page information is stored on computer disc.

A real-time program handles all requests for data from any source and provides the user with busy/free device information while causing the data to be transferred from or to the global area of memory.

4.12.7   *Input/output*

There is a variety of different types of Input and Output channel. Some are discrete, others are analogue or even serial block transfers of information. The control of the channels and scheduling of data transfers from/to the cockpit is the task of the Input/Output scheduler. This determines the type, size and frequency of transfer of data to the electronic interface. Most elements in the simulator Input/Output Interfacing System can be resolved into basic categories:

synchronous outputs – are used to drive instruments of the type having synchro torque receivers or control transformers, and are obtained by multiplying ac reference signals with an amplitude defined by a digital data word

analogue outputs – are used to convert digital words into analogue voltages to drive moving coil type instruments, etc.

analogue inputs – are used to convert analogue voltages representing control lever positions, etc., into digital words

discrete outputs – are used to convert digital logical data into voltages/currents for driving relays, lighting indicating lamps, etc.

discrete inputs – are used to convert the state of the switches, circuit breakers, etc., into digital logic data

serial data inputs/outputs – are used to interface the computer in simulations of some modern aircraft which themselves employ digital techniques of communication between equipment and instrument/displays.

4.13 The use of microprocessors in flight simulation

With the advent of first the 16 bit, and more recently the 32 bit microprocessor, the choices available to the simulator electronics design engineer have increased dramatically. Instead of the traditional, and limited, serial method of computing, using Single or Dual CPU configurations, it is now both possible and cost effective to utilise true parallel processing imbedded microprocessors. These can be chosen from the wide range available to suit particular applications where their high speed, power and invisibility to the user can be put to the best effect.

One example of such is the engine thrust lever characteristics where a simulation avoiding the use of expensive aircraft parts (the autothrottle drive mechanism) can be very cost effective. The feel forces experienced by the pilot during manual lever operation, the actuation of the thrust levers under autothrottle control, the baulks to prevent premature scheduling of thrust reversers; all these features can be conveniently handled by a microprocessor-based digital control loading system in a similar way to the simulation of the primary flying controls.

A mathematical model of the aircraft system is deduced, which is then transformed into a real-time software simulation program. This is 'burned-into' the memory chips which are accessed by the microprocessor. Communication between the microprocessor and the host computer is generally achieved using a high speed serial data link and by this means control of the functions computed within the micro can be exercised. Thrust lever position is thereby furnished to the engines model within the host computer and, when required, autothrottle rates can be commanded from the host computer. By means of analogue to digital converters, force and position transducers can be read by the microprocessor as inputs to the simulation model.

Then, within a very short frame time of perhaps 1 to 2 ms, a new output is computed and injected into a digital to analogue converter to schedule a hydraulic servo jack to update the thrust lever position. The high computational rates possible assure a smooth lever movement for the pilot.

5

Structures and cockpit systems

5.1   Introduction

The requirements for a flight simulator cockpit are well illustrated by the US Federal Aviation Administration Advisory Circular 120-40 (FAA 1983) and by the UK Civil Aviation Authority document CAP-453 (CAA). These call for the flight deck to represent accurately a specific aircraft type and model, to contain an on-board instructor station and to have provision for observers. Although these requirements specifically apply to simulators for commercial airline operation, the accurate representation of the flight deck is also required for military simulators. Fighter type simulators do not require an on-board instructor station or provision for observers, therefore other means have to be found to monitor closely trainee performance.

This chapter discusses the design, material and build of the cockpit structure, examines the electromechanical services required within the structure and the means of interfacing other simulator systems.

5.2   Mechanical and structural constraints

The simulator cockpit (or flight deck) is mounted on a platform that provides an interface to the motion system and contains services such as power, air conditioning and hydraulics as well as routing for power and signal cables. Also mounted on the base frame is the visual display system, if one is fitted.

The platform is generally a rigid space frame construction optimised for minimum weight and maximum stiffness. The structure itself is subject to the forces from the motion system (which on most present day simulators occur at three points), to the forces from the pilot's controls and is also subject to structural loading from the visual system. The highest level (Phase III) in the requirements (FAA 1983) calls for characteristic buffet and turbulence motion to be felt on the simulator flight deck. This means that the platform and all the structures, including the visual system, that are attached to it, should have minimum natural resonant frequency greater than the buffet and turbulence frequencies expected; in practice this is likely to be not less than 20 Hz.

In order to ease the manufacturing task the control loading units are often mounted in a separate module which can be installed into the platform after population and initial set-up. Easy access to the control loading units must be provided for maintenance purposes. In the past it was possible to allow access from either forward of the units or from below the units. Because of the very wide angle visual systems being developed access now has to be from below. This results in a large open area in the platform thereby complicating the space frame structure.

Further structural complication is caused by the necessity to provide routing for the cockpit air conditioning system, hydraulics for the control loading system and power for on-board electronics and signals to and from off-board computing and visual systems.

Another constraint faced by the designer is the necessity to provide flexibility of shipping options. In general the maximum size of any component of the simulator should be limited to fit into a Boeing 747 aircraft. For certain destinations smaller sizes are required to fit into available aircraft. Normally if air shipment restrictions are met there is then no difficulty with land or sea transportation.

The structural members of the platform are sized to meet a minimum stress reserve of 4, based on material yield properties when loaded to the maximum payload and subject to the maximum dynamic capability of the system.

Fig. 5.1. Computer model of baseframe.

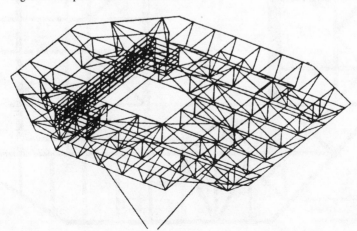

The design of the platform is therefore a compromise taking account of the maximum all up weight, the deflection sensitivity of loads, such as visual system mirrors, ease of manufacturing by using standard assemblies and accessibility of equipment for ease of maintenance.

The ready availability of finite element modelling software enables an optimum space frame to be designed. The ease of 'what-if' type calculations then enables the best compromise to be achieved. Fig. 5.1 shows a finite element model of a typical platform with equal motion leg excursions. The 'hole' for the control loading module can be seen.

5.2.1   Platform services

Packaged within the platform spaceframe are services such as air conditioning, hydraulics, power and signal cables; see Fig. 5.2.

Air conditioning in the simulator is required to:

(1)   perform generally as in the actual aircraft. However, it does not necessarily have to reproduce the full cockpit temperature range or

Fig. 5.2. Typical platform showing services.

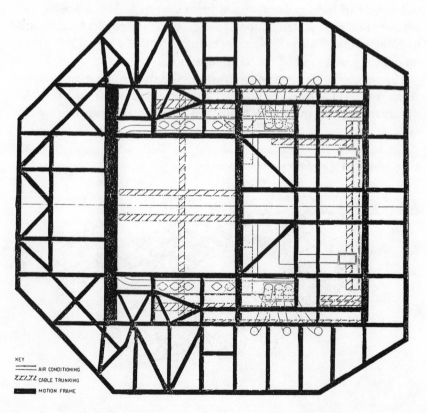

KEY

═══════ AIR CONDITIONING

ⲌⲌⲌⳐⳐ CABLE TRUNKING

▓▓▓▓ MOTION FRAME

the same rate of change of temperature. Similarly changes in cabin pressure are not normally simulated

(2)   provide cooling for on-board electronic equipment and instruments

(3)   in the case of simulators with on-board instructor stations and observers, to provide a comfortable working environment

The air conditioning plant is normally located off-board and the cooled air at low pressure but large volume is ducted into a plenum chamber in the platform. The air is distributed from this point through ducting sized to meet the operating conditions. The air distribution to the cockpit and instructor/operator areas is passed over heaters controlled by thermostats or simulated inputs. In addition, the air flow for the cockpit has to be controlled as a function of aircraft system operation i.e. is ground power connected or is the aircraft's auxiliary power unit operating. This is normally implemented by flap valves or boost fans.

Air for the on-board electronic equipment does not, of course, need heating. In the case of the simulator electronics the air is simply ducted to the cabinets. The layout of the equipment within the cabinet is designed to facilitate the flow of air. If necessary exhaust fans are provided to improve the air flow and hence reduce the maximum temperature rise within the cabinet.

Increasingly complex aircraft avionics equipment is being mounted within the simulator cockpit area. This equipment requires air at high pressure with a restricted temperature range. This necessitates the provision of special plenum chambers and air handlers to convert the low pressure air to the high pressure levels required by the aircraft equipment.

Airflow detectors are provided within the air conditioning system to switch off power to the electronic equipment in the case of air conditioning failure. Smoke detection systems are also provided to switch off the air conditioning system and simulator power if smoke is detected.

Hydraulic power is required for the operation of the control loading systems. Typically 0.5 in (12.5 mm) flexible pipe brings the hydraulic fluid into the baseframe. The fluid is then distributed by stainless steel piping to the various control elements. All hydraulic fittings are capable of leaking and must therefore be accessible for maintenance purposes. This means that the routing for the hydraulics must avoid locations where oil leakage can corrode cables or enter electronic equipment. At the same time maintenance access must be ensured including space for the operation of tools.

Where possible, power is routed as either 3 phase or single phase ac to the power supplies located within the structure. The power cables are routed to avoid close contact with the signal cables to avoid any noise pick-up. Standard engineering practices developed over many years in the electrical and electronic industries are used. The power supplies themselves are either located within the baseframe or within the electronic cabinets depending to a large extent on space availability and the size and weight of the power supplies themselves and the final dc distribution requirements. Linear bulk supplies are physically large and heavy. As a result it is preferable to mount them at as low a level as possible to minimise impact on payload centre of gravity. The most obvious position is therefore within the motion baseframe and this has been used for many simulators. Because of the weight of the supplies, occupational safety dictates two man removal and often special access platforms for maintenance. The use of switch mode power supplies is becoming more common and the greater efficiency of these units means more power is available in less volume and with less weight, thus easing the packaging problem. The normal dc voltages distributed through the cockpit structure are $+5$ V, $+/-28$ V and $+/-15$ V. The $+5$ V supplies are often very high current (200 A or more). To avoid power drops, physically large distribution is required either in the form of copper bus-bars or by the use of insulated braid of equivalent cross-section.

Aircraft equipment requires 400 Hz supplies either at 115 V or at 26 V. Aircraft electronic units requiring 400 Hz are normally driven from a solid state power supply. Heavy duty aircraft electromechanical equipment such as crew seats are very often driven from rotating alternators.

Separate power also has to be provided for:

(1)   Lighting. This is in addition to the normal cockpit lighting and is required for entry and exit from the cockpit and for maintenance activities.

(2)   Emergency lighting. This is continuously on charge and operates automatically in the event of a power failure.

(3)   Utility sockets. This is a separate supply for use with test equipment. Generally this uses the normal national supply voltage and frequency with the correct type of national sockets.

Signal cables are as far as possible manufactured from flat ribbon cable using insulation displacement connectors to minimise manufacturing labour. However, aircraft equipment, such as instruments, uses different types of normally circular connectors. Consequently transition connectors, strategically located to allow easy access, are required. Physically, signal cables are routed in troughs to separate them from power cables. Cables to individual aircraft instruments or avionic equipments must have sufficient slack to allow easy removal of the instrument or panel from within the cockpit.

The multiplicity of signals and power supplies, both ac and dc, requires a sophisticated grounding system using isolated heavy duty bus-bars where possible. Ground/earth loops are avoided by adherence to good design and wiring practices. Grounds are kept isolated as far as possible back to the facility earth point. The typical types of ground are:

> D-Ground: for digital signals. Normally reference for 5 V supplies

Fig. 5.3. Typical grounding configuration.

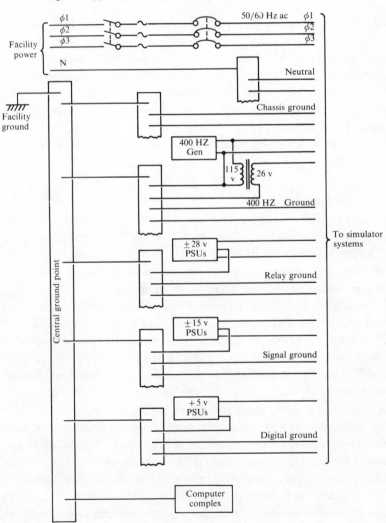

S-Ground: for analogue signals. Normally reference for 15 V supplies

R-Ground: for lamps and relays and other heavy duty switching type signals. Normally reference for 28 V supplies

C-Ground: safety ground for metal surfaces

400 Hz-
Ground: separate ground for 400 Hz supplies

Fig. 5.3 shows an example.

Identification is provided for all cables, cable bundles, connectors and single terminations. The exact methods are determined by the customer's requirements.

The services to the motion platform enter via an umbilical from the ground which, due to the physical resemblance, is known as a waterfall.

The waterfall must be located in a position where there is no possibility of contamination by fluid leakage from other simulator equipment.

To minimise flexing of the pipes and cables the ideal entry point to the motion platform should be as near to the motion centroid as possible. Attachment of the cables to the simulator is generally through wire mesh sleeves (known as 'Chinese Fingers') to take the weight and relieve the stress on the cables.

5.3   Cockpit systems

The simulator cockpit is the baseframe and, viewed from the interior, it is an exact replica of the aircraft cockpit. The actual structure of the cockpit is determined by the type of aircraft being simulated.

For single and two seat fighter type aircraft a metal skinned structure is often used with 'as-aircraft' canopies and canopy rails.

For commercial aircraft and also for bomber and transport type military aircraft glass reinforced plastic is mainly used for the cockpit shell.

The cockpit structure is mounted onto the baseframe. For ease of manufacture and also to reduce overall costs the baseframe is of a standard design. The cockpit is interfaced to the baseframe via a unique interface frame which includes facilities for the installation of aircraft unique equipment such as seat rails and the shell itself.

For large aircraft the instructors are carried on-board the motion platform whereas for fighter aircraft the instructors are located off-board. Therefore for large aircraft the cockpit structure has to be merged into an enclosure to house the instructors. For two crew aircraft such as Boeing 737 or 767, the transition between the cockpit area and the instructor's area can take place immediately aft of the crew seats and the centre pedestal.

However, for three (or more) crew aircraft such as the McDonnell Douglas DC-10, Boeing 747 or Lockheed C-130 the separation between the third crew member and the instructor is not so clear cut. In fact the pilot instructor often sits alongside the flight engineer.

Access to these cockpits is via a door in the rear. For single and tandem seat aircraft access is via the canopy opening. When visual systems are installed, the positioning of the optical system can interfere with the canopy opening. The designer then has to make the choice of either modifying the canopy opening or moving the optics when access is required. In order to maintain the illusion of being in an aircraft, movement of the optics should be the preferred solution rather than modification of the canopy. However, this can lead to increasing complexity with consequent impact on reliability and maintainability. In addition, the set-up of the optical system can be impaired with consequent degradation of visual scene quality or emergency egress from the cockpit can take too long. Therefore, as with the design of the motion platform, engineering compromises have to be made.

The use of aircraft parts in the simulator is an area of much discussion between the simulator manufacturer and the simulator user. The decision on whether or not to use actual aircraft parts is normally made on cost grounds. Aircraft parts are expensive and are generally designed either to operate in a harsher environment than the simulator or to operate for a shorter time period than in the simulator. However, certain uses of 'as-aircraft' parts do not lead to conflict between the parties. These are:

(1)   Aircraft items that must appear identical in the simulator such as throttle levers, trim wheels etc

(2)   Aircraft pressure operated instruments. Since air pressure systems are not normally present in the simulator these pressure driven instruments have to be replaced by electrically driven instruments.

5.3.1   *Instruments*
For aircraft instruments three choices are available. These are:

(1)   'as-aircraft'

(2)   modified

(3)   simulated

The use of 'as-aircraft' instruments has the advantage of the instruments being 100% representative of the aircraft. In addition support can be shared with the aircraft. Many airline and military users do in fact treat the simulator as another aircraft for spares or maintenance purposes. As far as the simulator manufacturer is concerned the use of 'as-aircraft' instruments has a number of disadvantages. Instruments in the simulator are subject to high cyclic usage which is liable to result in a lower mean time between failure with consequent impact on simulator availability. Many aircraft instruments have long delivery times as do their mating connectors. This leads to increased simulator delivery times with consequent cost increases. The 'as-aircraft' instruments are designed to work with aircraft sensors and in unique configurations thus giving rise to costly unique drive interfaces. In addition the instruments are liable to use the chassis as an earth connection thereby introducing ground reference problems into the simulator with possible impact on performance.

Some of the problems outlined above can be overcome by modifying 'as-aircraft' instruments to aid the simulation task. Changes can be minimal, such as the modification to the slip ball on the ADI, or extensive. In the latter case circuits within the instrument may be bypassed or significantly rewired. The more the modification of course the better it may be totally to simulate the instrument.

Instruments can be simulated by the simulator manufacturer or by the instrument vendor. In the latter case the instrument interfacing problem is eased but different vendors do have their own preferred solutions. If the simulator manufacturer produces the instrument then it is of course designed for simulator use and uses more robust parts to permit higher cyclic usage. Also the electrical interface is matched to the available simulator signals giving, for example, maximum use of available resolution. Standard cabling and connector interfaces can be used and extensive use made of a limited range of standard parts such as motors, amplifiers and potentiometers. Fig. 5.4 shows a typical simulated instrument manufac-

Fig. 5.4. Simulated instrument.

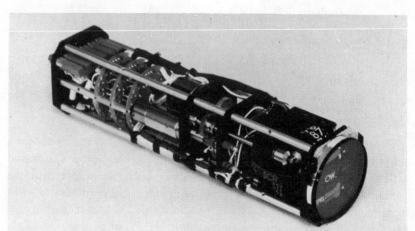

tured by the Singer Company; this uses the 'In-line Motor Potentiometer' (IMP) system that results in a low-torque device using a plastic conductive throughshaft potentiometer in line with a high powered motor producing up to 340 degrees of free rotation. This mechanism can be adopted for a wide range of single drive simulated instruments. There are of course disadvantages to the use of simulated instruments. For instance logistic support in common with the aircraft is no longer possible. Also the incorporation of the simulated drive can make the simulated instrument longer than its aircraft equivalent thereby making installation more difficult.

Where simulated instruments are used they are predominantly driven by analogue signals. The actual drive types can be split into the following six classes:

Class 1: *dc servo instrument, limited rotation*

With this class of drive the pointer is driven by a standard dc motor via a gearbox. An analogue signal provides the control input. A servo amplifier compares this signal with that provided by an answer potentiometer driven by the gearbox. The resulting error signal controls the motor.

This class of drive is used where the required pointer travel is greater than that provided by a dc moving coil (250 degrees) or where high torque is required. Typically engine tachometers and fuel indicators use this class of drive.

Class 2: *moving coil instruments/IMP drive*

These two drive systems are classified under one heading, the moving coil system can be superseded by the IMP driven instrument. The potentiometer characteristics of the IMP system permit simulation of angular ranges up to a maximum of 340 degrees and simple amplifier adjustments enable unique instrument characteristics e.g. overswing, to be faithfully represented.

Class 3: *dc synchro repeater instrument*

A dc synchro repeater of a standard type is used for this class of instrument drive.

This class of drive is used where pointers are required to be capable of continuous rotation and there is no requirement for high torque.

Typically, instruments such as radio magnetic indicators (RMI), standby compasses and distance measuring equipment (DME) readouts are in this class.

Two solid-state, voltage-follower amplifiers are packaged in the instrument case to provide the necessary power amplification. These permit the instrument to be driven directly by two D/A converters in the real-time interface system.

Class 4: *dual speed dc servo instruments*

This class of instrument is driven by a standard dc motor (as class 1) coupled to the pointer via a gearbox with two answer potentiometers. Coarse positioning of the pointer is provided by a multiturn potentiometer, and fine positioning by a sine–cosine potentiometer.

Instrument drives of this class are used on instruments having a limited multiturn capacity and a high accuracy requirement.

Typically, altimeters use this class of drive system.

Class 5: *dc sine–cosine servo instruments*

This class of instrument is driven by a standard dc motor (as Class 1 and 4) coupled via a gearbox to the pointer or dial to be driven with a sine–cosine potentiometer (as Class 4) providing positional feedback.

Instrument drives of this class are used where pointers or dials are to be capable of continuous rotation and a high torque is required.

Typically, instruments such as attitude indicators, cards of RMIs and HSIs use this class of drive.

Class 6: *electromagnetic counters*

This class of instrument is driven by digital outputs. Each number on the counter wheel has a corresponding terminal within the instrument.

Energization of a terminal rotates the wheel to the associated number by means of an electromagnet.

Typical counter instruments are DME indicators, course indicators and altimeter readouts.

Crt displays

With the increased use of crt displays in aircraft – the glass cockpit – and other sophisticated systems, the use of 'as-aircraft' avionics equipments will increase in the simulator. This is due to:

(1)   the necessity of having identical software in the simulator and in the aircraft; an FAA requirement

(2)   the sophisticated graphics of the cockpit displays resulting from the major investment by the avionics manufacturer that cannot be matched by the simulator manufacturer within a reasonable time frame

(3)   the use of standardised interfaces (MIL STD 1553B (1980) or ARINC (1981))

In the future, the ideal simulated instrument will have the identical physical appearance of the real part, contain its own interface electronics and will require a minimum number of signal and power connections possibly to a single common bus standard.

5.4   Electronic packaging

The simulator cockpit contains not only the aircraft equipment, but also the electronics required to interface to the simulator computer system. In addition provision has to be made for special simulator systems such as communications, aural cues and lighting system. Also for large simulators provision has to be made for an instructor's station.

The simulator designer has therefore to determine exactly what has to be installed and what are the space conflicts with other equipments and personnel such as instructors and observers. Besides the actual electronic hardware, provision has to be made for the cabling and connectors and for power supply units. The overall power supply philosophy has to be determined. For instance, should large bulk supplies be centrally located with power distributed via heavy duty cabling to avoid voltage drops, or

Fig. 5.5. Electronic packaging in platform module.

should smaller power supplies be distributed throughout the electronics. In many cases the simulation manufacturer will provide a bulk solution for one type of simulator and a distributed solution for another type.

Maintenance access is a prime design consideration as is standardisation of cabling with easy breakdown for shipping. An approach adopted for fighter type aircraft has all the electronics packaged behind the cockpit in a free-standing rack system with swing-out card bins; an example is shown in Fig. 5.5. For large commercial and military simulators an approach adopted by one major simulator manufacturer has been to locate the electronics outside the cockpit shell with access from the platform; an example of such a layout is shown in Fig. 5.6.

The whole simulator is then enclosed in a housing to maintain the simulator environment constant as shown in Fig. 5.7. The present trend to large, wide field-of-view, visual systems means that this approach will no longer be feasible. Space will therefore have to be provided in the area presently reserved for instructors and observers. Thus an advance in one area of simulation has brought difficulties in another area.

The electronics are normally designed as stand-alone circuit boards. The size of these boards has been increasing from the 4 in. × 8 in. (100 mm × 200 mm) boards common in the 1960s to boards of approximately 9 in. ×

Fig. 5.6. Packaging (artist's impression).

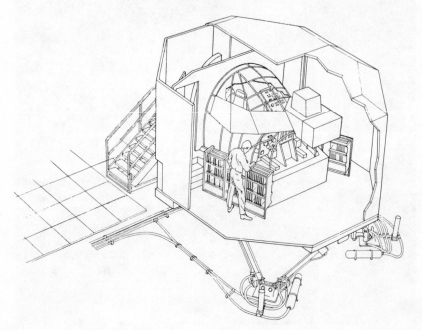

10 in. (225 mm × 250 mm). One of the reasons for this increase in size is the provision of standard diagnostic hardware on every board. The boards themselves can be fabricated either as printed circuits or wire-wrapped boards. The choice of whether to design as wire-wrap or as printed circuit depends on the quantity of boards to be built. However, with the use of Computer Aided Engineering and Design systems the choice of printed circuit can be justified at very low board counts. The use of printed circuits also means that standard diagnostic and test hardware can be readily included on the board and that the circuit board is totally independent of its location in the simulator. This can sometimes be a problem with wire-wrap boards since the coupling of signals can be different on supposedly identical boards.

With digital flight simulators an interface is required between the simulation model cycling in the computer and the aircraft equipment in the cockpit. All computers support interfaces to common peripherals such as line printers, bulk storage devices and terminals but no traditional computer has been marketed to interface to a peripheral as complex as a flight simulator. Some computer manufacturers do provide interfaces that are designed for the process control market which in general do not require

Fig. 5.7. AST simulator external view.

the special interfaces desired for the simulator or the resolution and low conversion times needed for real-time simulator interfacing.

Simulator manufacturers have traditionally made use of Direct Memory Access (DMA) channels, with their high, data-throughput capabilities, to transfer parameters between the simulator and the computer. Within the simulator each data source has to be interrogated in turn and all data outputs statiscised to provide continuous conditions. The hardware providing the interrogation/staticising action is typically referred to as the linkage. Embodied within the linkage is the signal conversion equipment to provide the final interface to the cockpit equipment. The main types of circuits used are:

(1)   analogue to digital converters (A/Ds or AIs) to input analogue signals

(2)   digital to analogue converters (D/As or AOs) to output analogue signals

(3)   discrete inputs (DIs) for input of switch states or other digital type signals

(4)   discrete outputs (DOs) which are open collector drives for lamps and relays etc.

In addition modern linkages are likely to include one or more of the following special features:

(1)   special cockpit interfaces such as

     ac synchro drives

     dc synchro drives

     ac resolver inputs

     variable reluctance drives

     Serial Data Channels (for MIL STD 1553

     or ARINC 429 for example)

(2)   conversion of floating point data into analogue variables and vice versa

(3)   packing and unpacking of logical variables (typically bytes, halfwords (16 bits) or words (32 bits) into discrete digital inputs and outputs.

The conversion and packing/unpacking hardware are generally located off-board and close to the computer hardware to minimise the length of the DMA interface cable. The remainder of the conversion equipment can be located on-board and is often connected by a serial data channel to minimise the cabling requirements.

As aircraft equipment has become more complex the quantity of linkage circuits has increased. However, with the advent of the 'glass cockpit' the mix of circuit types has started to change as shown in Fig. 5.8. A significant reduction in synchro drivers has occurred which is more than compensated by the increase in serial channels. In future more and more serial channels are likely with a further reduction in analogue requirements.

When selecting or designing a linkage for a simulator a number of aspects have to be taken into consideration, such as data throughput, maintenance and diagnostic capabilities and failure modes.

In the case of data throughput, what is the linkage update rate required and how does it compare with the frame time of the simulator computer? If the number of inputs or outputs of each type is large or the internal conversion time exceeds the capability of the DMA channels then extra linkage channels would be required. If future expansion is required then this must also be taken into account when determining data throughput.

Maintenance and diagnostics features are becoming more and more important. Maintenance capability must be designed into the system from the beginning. For instance, if the system has cables terminated on the front of the circuit boards then access for maintenance is reduced. Diagnostic features in the linkage also enhance maintenance capability. Ideally the features should include wrap around test of A/D driving D/As etc. to prove

Fig. 5.8. I/O usage by type.

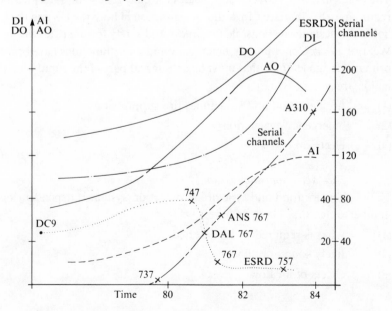

operation right at the output. Means should also be provided to report status via the computer either on demand by the operator or continuously if a 'hard' failure has occurred.

The packaging of the electronics and its associated cabling has to take into account the effect of radio frequency interference on the operation of the simulator and the operation of the simulator on nearby equipment. Simulators are often located on airfields in the vicinity of high powered radar systems which can induce signals in the simulator unless adequate precautions are taken. Military simulators have had to ensure that no radiated emissions could cause a breach of security. During recent years, government regulations in many parts of the world have defined acceptable emission levels. These regulations can be represented by the US Federal Communication Commission Rules Part 15 (FCC 1980). Simulators are defined as industrial equipment and therefore have to meet the FCC Regulations for Class A Equipment. This defines the maximum field strength radiated from the equipment and the maximum voltage that the equipment can feed back to the power lines and is applicable to all equipments that generate timing signals or pulses at frequencies equal to or greater than 10 kHz.

5.5   Emergency and safety features

Flight simulators are ground training devices and as such need to meet environmental and occupational safety and health requirements. The user's insurance cover also specifies the emergency and safety features required in the simulator. In addition legislation in both the US and the UK holds the designer responsible for fitness and safety for use (Cartwright & Wearne, 1984). Consequently, emergency and safety functions have become primary design features that must be an integral part of the simulator. The major areas are:

(1)   fire detection and sometimes fire suppression

(2)   emergency evacuation

(3)   electrical codes

5.5.1   *Fire detection/suppression*

The location and type of fire detection system is dependent on a number of factors. These include:

(1)   what is required to be detected

(2)   likely sources of fires

(3)   effects of air flow

(4)   power requirements

(5)   training requirements

Local fire regulations can determine the choice of sensors as well as the types of combustion sources to be detected. For instance, are elevated temperatures to be detected, or a rate of rise of temperature, or the visible products of combustion, or the invisible products of combustion, or any combination of these? Airflows also have a significant impact on both the choice and location of sensors. If the sensors are in areas subject to airflow, will the airflow increase the perceived ion density or will it dilute the particle content thus negating the effectiveness of the sensors? If the sensors can cope with these factors what will be the effect if the airflow is off or if the sensors are located in air stagnant areas?

The sensors, whether smoke or heat detectors or a combination, should give a warning to personnel in the cockpit and initiate an emergency power off. If a fire suppression system is fitted then this should also be automatically initiated. Many modern simulators have smoke generation systems which are activated by certain simulator malfunctions. In these cases the smoke detectors have to be deactivated.

The latest fire suppression systems make use of the gas Halon 1301. Halon 1301 is one of the group of halogenated hydrocarbons and is often called bromotrifluoromethane. It is a colourless, odourless gas which acts by breaking the combustion chain reaction (NFPA 1984).

The fire suppression system consists of a number of containers of the Halon which discharge through nozzles to release an adequate quantity of the gas into the protected area. The fire detection system automatically activates the discharge of the Halon which totally floods the cockpit and the area around the motion system. For the type of fire expected in a simulator a 5% concentration of Halon is required and this may have to be maintained for up to 10 minutes.

Although Halon is a low toxicity gas the manufacturers recommend that for the concentration used in the simulator, personnel evacuate the area within 1 minute.

Because of the specialised nature of fire suppression systems and the possible legal implications the design and installation of fire suppression systems is most often carried out by professionals in the field.

5.5.2   *Emergency evacuation*

Evacuation from the simulator should be possible in approximately 30 seconds in the event of a fire or the operation of any fire suppression system fitted. Generally speaking 10 seconds are required for the motion to settle and a further 10 seconds for the stairs or drawbridge to reach the level of the simulator platform. Certain regulatory authorities require more than one exit from the cockpit area particularly for wide-body commercial simulators. This is either provided by a frangible panel in the instructors' area or by allowing exit through one of the cockpit windows. Also some alternative way of leaving the motion platform has to be provided and this is often by means of a rope ladder stowed on the platform.

5.5.3   *Electrical codes*

The power distribution system of the simulator is designed in accordance with the electrical codes of the using location. All simulator manufacturers design to their national electrical codes and make modifications to suit the unique requirements if any, for a specific simulator.

Wiring practices conform to an internationally recognised standard typically the National Electric Code of the United States of America of the IEE Wiring Regulations for Great Britain.

6

Motion systems

6.1   Introduction

Aircraft have motion, and all simulators impose on the pilot cues of the motion that result from his control inputs to the simulated aircraft. Some of the aeroplane motions aid the pilot in stabilizing and manoeuvring the aircraft by confirming that the response of the aircraft matches his internal model of expected response. Others alert him to aircraft system failures. Still other aeroplane motions, such as those resulting from vibration or turbulence, make the pilot's task more difficult. In many simulators the only indication to the pilot of the motion of the simulated aircraft is through visual cues provided by instruments or by the simulated view of the outside world through the cockpit windows. However, experimental evidence shows that the method by which the pilot flies an aircraft is affected by motion detected by other human senses (Douvillier *et al.* 1960; Huddleston & Rolfe 1971; Levison & Junker 1978; Perry & Naish 1964) to a degree dependent on the aircraft, the task, and the manoeuvre. For this reason these senses are stimulated in some simulators by motion of the cockpit or other methods which affect directly the human vestibular and proprioceptive sensors. Systematic investigation and modelling of the human motion perception system and of pilot response to these perceptions have resulted in methods of choosing an appropriate motion-cue-producing system for training simulators and for engineering simulators used in aircraft research and development.

6.2   Motion sensing and perception and their modelling

The body uses a variety of sensors in the detection of motion, and the information detected by these sensors is processed by the body in arriving at a perception of the motion. The eyes are the most obvious sensors, detecting cues of motion in an aircraft or a flight simulator both from the cockpit instruments and from the pilot's view of the world outside the cockpit. While the visual cues of motion are probably the most important low frequency cues (Woomer & Williams 1978), they will be treated in this chapter only as they relate to the tactile sensors and the proprioceptive sensors, the subcutaneous sensors that respond to stimuli produced inside the body. These proprioceptive sensors include those associated with the vestibular system, the joints, the muscles, and the internal organs.

The vestibular system, the non-auditory portion of the inner ear, is an important and sensitive sensor of motion and position. The semicircular canals are the rotational motion sensors, and they act as damped angular accelerometers. There are three such canals in each ear which can detect angular acceleration in laboratory experiments as low as 0.1 degree/s^2 (Hosman & van der Vaart 1980; Meiry 1966). The otoliths of the inner ear are the linear motion sensors which detect specific force, the external or non-gravity force acting on the body. Their threshold has been measured as low as 0.02 m/s^2. Static body tilt is sensed as a change in the direction of the specific force vector, indistinguishable from a change caused by linear acceleration. The otoliths can detect a tilt of about 2 degrees. Both sets of these organs detect head motion, whether it is caused by motion of the whole body or motion of the head with respect to the rest of the body. The outputs of these and all other sensors appear as neurological signals, afferent firing rates, which are transmitted through the nervous system to the brain. Additional proprioceptive sensors are located throughout the body. Muscle extensions are sensed, both those that arise from movement of the head and limbs relative to the torso and those of the internal organs. In addition, the tactile sensors detect pressure or force applied to the external surface of the body. While, from a qualitative viewpoint, the responses of these nonvestibular sensors are somewhat more self-evident, they have not been studied or modelled as completely as those of the vestibular system (Borah, Young & Curry 1979; Gum 1972).

The study of the results of a number of manned aircraft simulations by Young (1967) has led to the development of a systems approach to human perception of orientation and motion with particular application to flight. Using this approach, the vestibular sensors have been modelled as mechanical systems (Young *et al.* 1969; Young *et al.* 1973), the constants of which have been validated by comparison with results of experiments of human perception. Ormsby (1974) has modelled the vestibular apparatus to show the transfer function between the physical stimulus and the afferent firing rate (expressed in terms of stimulus units) as: semi-circular canal

$$\frac{y_{\text{scc}}}{\phi} = \frac{0.07s^3(s+50)}{(s+0.05)(s+0.03)}$$

$(\text{deg/s}^2)/(\text{deg/s}^2)$

otolith

$$\frac{y_{\text{oto}}}{f} = \frac{2.02(s+0.1)}{(s+0.2)}$$

For these dynamics, velocity thresholds are about 2.5 degree/s and 0.2 m/s.

In addition to the modelling of these purely physiological sensors, the complete model of perception must include the psychophysical aspects, the mental processing which each individual, depending on his environment, his mind set, and his experience, uses to convert the output of the physiological sensors into his perception of motion (Borah, Young & Curry 1979). Their model of human perception includes this processing, and its basic structure is shown in Fig. 6.1.

The psychophysical processing of the sensor signals by the central nervous system is modelled as an optimal estimator which can be reduced to a steady-state Kalman filter. While the nature of this filter can be verified by experiment, the filter constants are determined both physiologically and by the wide possible range of experience and training of each individual. By computer simulation, Borah has verified the model, using a version which responds only to angular and translational velocities and in which the visual sensor model includes only vehicle out-the-window peripheral cues. The results match a number of experimentally-known human perceptions

Fig. 6.1. Basic structure of motion and orientation perception model (from Borah, Young & Curry 1979).

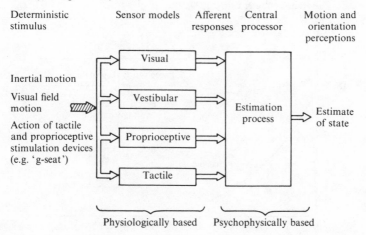

of motion, particularly in yaw where the greatest body of experimental results exists. Fig. 6.2 shows the result of one of these simulations which tested the model response in yaw and shows the relative response as a function of time to the visual cues of motion, the vestibular cues of motion and the combination of the two.

The experience and training of the individual, the frequency response of the various sensors, and the simultaneous reception of signals from different sensors will greatly alter the individual's perception of motion. The thresholds of perception of linear and rotational motion will be increased by pilot workload (Hosman & van der Vaart 1980), superimposed vibration, and other distractions. The spectrum where the various sensors are most sensitive ranges from vision at the lowest frequency through the otoliths, the semicircular canals, and the other proprioceptors to the tactile sensors at the upper end of the frequency range. The implication of this spectrum is that stimulation of the non-visual sensors above their threshold values will result in the earliest recognition of motion, the important reason for stimulating these sensors in some aircraft simulations.

6.3   Production of motion cues

6.3.1   *Motion platform design*
Motion systems are generally characterized by their excursion, velocity, and acceleration in their several degrees of freedom. With few exceptions motion systems are hydraulically operated to take advantage of the ruggedness and small mass and physical size of the hydraulic

Fig. 6.2. Perception model response to visual and inertial cues in yaw (after Borah, Young & Curry 1979).

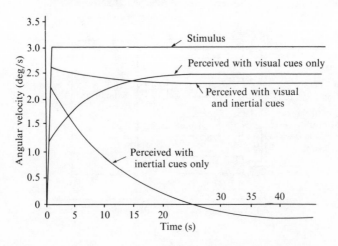

components. While platform position is the basic reference variable, the servomechanisms are often designed to respond primarily to velocity or acceleration of the reference to increase response and accuracy (Mills 1967). Because the bandwidth of rigid airframe dynamics of the simulated vehicle and of pilot control is limited to 2.0 Hz or less, approximately flat amplitude ratio response is required over this range, if flying qualities are to be reproduced. If airframe first order bending effects on the pilot are to be simulated, the frequency range must be extended to about 5.0 Hz.

It is important that the motion platform should not generate motion cues that are uncharacteristic of the aircraft. Any commanded motion will generate some spurious accelerations arising from structural or servomechanism resonances, friction, and other system deficiencies. The system must be designed for lowest structural resonances to be well out of the frequency band of interest while at the same time making the structure low enough in mass to minimize power requirements (Engelbert *et al.* 1976; Mills 1967). Hydrostatic bearings will minimize friction (Baret 1978; Koevermans & Jansen 1976), and careful attention to hydraulic servovalve design (Lacroix 1979) can decrease response time and minimum commanded movement.

Non-linearities and cross-coupling can directly decrease the useable range of the motion platform by decreasing the fidelity and so the realism of the commanded motion. Spurious acceleration and other system deficiencies can only be minimised, not eliminated. Noise ratio is one measure of quality and is defined (when the system is driven by a sine wave) as: the

Fig. 6.3. Typical motion platform system and operational limits (after AGARD (1979)).

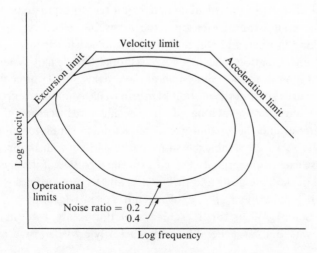

standard deviation of the output acceleration from the fundamental output divided by the standard deviation of this fundamental output.

Fig. 6.3 shows typical characteristics for noise ratio. This noise normally limits the useable platform excursions and velocities to somewhat less than their maximum values. Note that acceleration noise is produced not only along or about the axis considered, but that parasitic noise is produced along or about other axes by cross-coupling effects. AGARD (1979) discusses these deficiencies and describes methods of measuring them.

6.3.2   *Motion platform drive signals*

Signals representative of the acceleration and the attitude of the simulated aircraft must be converted to signals which will drive the motion platform. These signals must drive the platform in a manner to reproduce at least a portion of the spectrum of the forces which would be acting as the pilot in the cockpit of the simulated aircraft. Except for very limited test conditions, such as helicopter hover, no motion platform of reasonable size can move the simulator cockpit in a manner which provides a one-to-one correspondence with the forces the pilot would experience in the simulated aircraft. At best, the motion platform can provide the simulator pilot with an acceptable perception of these inertial cues of motion. The objective is to make the motion platform maintain in real-time an acceptable approximation of the rotational accelerations and of the specific force vector at the pilot's position in the simulated aircraft. To generate these forces and accelerations while keeping the motion platform within its excursion limits, the signals representing the aircraft accelerations must be modified to drive best the motion platform. This signal treatment includes attenuation, limiting, and washout of the calculated aircraft forces to simulate best the forces on the pilot without exceeding the allowable excursions of the platform; tilting in pitch and roll to provide sustained linear forces; and coordinate transformation to change force vectors from the axis system of the simulated aircraft to the axis system of the moving simulator cockpit.

Washout is used to limit the motion platform to its available excursions while maintaining best matching of the specific force vectors of the simulated aircraft and the simulator cockpit. Training simulator motion platforms have relatively small excursions, particularly in translation, and they use these excursions primarily to control the *direction* of the specific force vector, to simulate gusts and turbulence, and to provide cues which confirm to the pilot the response of aircraft systems (gear in transit, touchdown bump). When small excursion platforms are used in this manner, the washout can generally be satisfactorily provided by a high-pass, first-order filter between the signal that represents a given aircraft acceleration and the signal that commands the corresponding platform acceleration. This filter removes from the platform command, the sustained excursion which would soon drive the platform toward its excursion limit, where the resultant deceleration would provide an unrealistic cue. Some pure attenuation of the commanded force is acceptable to the simulator pilot (Parrish 1978; Sinacori 1973; Young, Curry & Albery 1976), and it will be required in almost all manoeuvres to limit motion platform command to the available travel. Sustained forward (surge) and lateral (sway) accelerations can be reproduced by pitch and roll motion of the platform in the manner described below.

When platform motion is used in aircraft handling qualities research and development, more elaborate washout schemes are justified to simulate better the aircraft response to control inputs. The most usual method is to pass each signal representative of the aircraft accelerations through a high-pass filter of at least the second order, although higher order filters may be used, and good results have been obtained from a non-linear washout scheme which eliminates a false cue introduced by the more usual methods (Parrish *et al.* 1975; Parrish & Martine 1976). Formal experiments to determine acceptable attenuation and phase lag of the force vector are necessarily limited in scope (Jex *et al.* 1979; van Gool 1978), but some indication of the range of values, based on the experience of Sinacori and Bray, is summarized in Fig. 6.4. This experience is the result of work on

Fig. 6.4. Subjective motion fidelity at one radian/second (from Sinacori 1977).

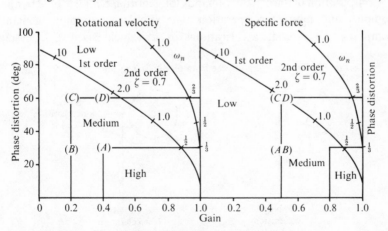

High: Motion sensations are close to those of visual flight
Medium: Motion sensation differences are noticeable but not objectionable
Low: Differences are noticeable and objectionable, loss of performance, disorientation

several motion platforms simulating many different vehicles (Bray 1972; Bray 1982; Sinacori 1977).

Residual tilt is a method of providing the sustained forces of translational acceleration by reorienting the pilot with respect to gravity. The gravity vector added to the force from the motion platform in sway yields a resultant vector at an angle to the pilot's vertical which can be matched in direction and approximated in magnitude by an off-vertical roll position of the platform as shown in Fig. 6.5. When both roll and sway motions are available in a motion platform to reproduce both the onset and the holding of a sustained side force, both are used. As a result roll acceleration can be kept low as the roll position is changed to produce a simulation of sustained side force. Pitch can of course be used with surge in a similar manner to simulate sustained surge forces.

The force resulting from using roll position as a method of producing sustained side force becomes a disadvantage when the roll platform motion is used to simulate aircraft roll acceleration. In a conventional fixed-wing aircraft, roll acceleration generally occurs in a coordinated aircraft manoeuvre which keeps the specific force vector small and oriented downward with respect to the pilot. To achieve a similar vector orientation in the simulator would require large lateral or sway excursions of the motion platform. The usual effective compromise is to attenuate strongly the roll motion. This compromise gives some cue of roll acceleration and keeps the misdirected specific force vector small. Generally, less attenuation is required in pitch and yaw, because high-frequency manoeuvre amplitudes in these modes are much less than those in roll.

Computation of the drive commands for a motion platform is generally a complex and multi-step operation. The simulator computer generally computes the aircraft accelerations in a vehicle-based axis system.

Fig. 6.5. Sway acceleration cue from residual tilt in roll.

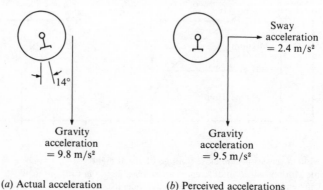

(*a*) Actual acceleration   (*b*) Perceived accelerations

However, the reference for the motion platform which the computer drives will be another axis system which is earth-based. Therefore, coordinate transformation is required to determine forces and accelerations in this earth-based system. Integrators can provide the motion platform servo system with the velocity and position commands which may be required by the machinery designer. Limiting of these commands in the simulator computer can provide additional safety to the simulator pilot from motion hard-over failures by providing software protection that is redundant to that provided in the platform hardware. Most motion platforms require the movement of more than one actuator to attain platform motion in a single degree of freedom. Therefore, additional conversion of signals from those representative of platform acceleration to those representative of actuator commands is necessary.

Fig. 6.6 is a simplified diagram showing one method of how the signals representing accelerations of the simulated aircraft roll and sway are transformed to motion platform drive signals. Detailed diagrams of the complete signal treatment and suggestions for setting optimum attenuations and filter frequencies can be found elsewhere (Gallagher 1970; Schmidt & Conrad 1970; Sinacori 1973 and Sinacori *et al.* 1977).

The choice of how to use the motion system in a given experiment or mode of simulated flight is equally important to the correct implementation and tuning of the motion drive signals. In engineering simulators good experiment design can ensure the use of the motion platform to provide inertial cues of motion most representative and acceptable to the pilot. For example, separating simulator investigation into separate tests of aircraft lateral and longitudinal flying qualities will allow separate optimum adjustment of the platform drive equations for the two tests and better

Fig. 6.6. Coordinated washout for a motion platform with sway and roll.

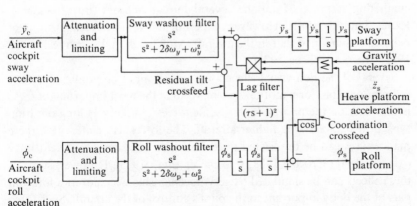

matching of aircraft specific force cues. Attempts to provide inertial cues of motion beyond the capabilities of the platform are often more confusing to the pilot than the complete absence of these cues (Cooper & Drinkwater 1971; Gilbert & Nguyen 1978; Lee 1971). The thoughtful extrapolation of data taken with realistic simulator motion cues is preferable to data taken with a pilot confused by conflicting cues.

Both training and engineering simulators can be programmed to use different motion platform drive logic and washout and attenuation settings in different flight phases (Ricard *et al.* 1980). Parrish reports that most satisfactory operation of the six-post synergistic platform in a fighter aircraft simulation is obtained by not using platform vertical motion and reserving the available actuator travel to reproduce better the other degrees of freedom (Parrish & Martin 1976). However, this same vertical travel has been used effectively for buffet, turbulence, and vibration simulation (Parrish & Martin 1976; Woomer & Williams 1978), and the drive logic can be changed to drive such a platform in the vertical direction only when these conditions occur.

6.3.3   *Non-platform motion-cueing devices*

The 'g-seat' is an effective device for producing some of the cues of aircraft motion by supplementing or replacing those produced by a motion platform. These devices are conventional aircraft seats built with a number of separate, inflatable cells. The cells are inflated and deflated in a properly controlled time sequence related to the motions of the simulated aircraft (Ashworth 1976; Kron & Kleinwaks 1978; Mathews & Martin 1978). Although there is little effect on the vestibular system, the other proprioceptive sensors and the tactile sensors are stimulated in a manner similar to aircraft flight. The optimum logic for driving the individual cells is not yet firmly established, since some seat movements can stimulate conflicting responses from the several human sensors (Bose, Leavy & Ramachandran 1981). However, experiments show influence on pilot behaviour similar to that in the aircraft (McKissick, Ashworth & Parrish 1983; Stark & Wilson 1973).

Several of the non-visual effects of motion cannot be produced by any conventional platform motion system. Among the most important of these is the effect on the pilot of high specific forces of relatively long duration typical of manoeuvring fighter aircraft. The forces associated with these manoeuvres can be duplicated over the entire body only by a centrifuge (Alberry 1981; Crosbie 1983; Crosbie & Elyth 1983). However, some of these effects can be simulated by servomechanisms which exert a force on part of the body important to the pilot's control of the aircraft. Such force

devices attached to the arm or to the head obviously influence pilot
response and have been shown to evoke responses by the simulator pilot
that correspond to his behaviour in flight (Ashworth & McKissick 1978;
Cardullo 1981; Kron, Cardullo & Young 1981).

The aircraft anti-'g suit', used in flight to overcome the blood-pooling
effects of high specific force, is often used in simulations of fighter aircraft. It
is of course not necessary for the suit to overcome physiological effects that
do not occur in the simulator. However, it produces a powerful sensation
that the pilot associates with these accelerations and thus can provide a
valuable cue of acceleration and change of acceleration (Staples 1978;
Stark & Wilson 1973; Stark 1976).

The high-frequency vertical forces associated with turbulence and
helicopter vibration are both fatiguing to the pilot and affect his ability to
read cockpit instruments and charts. Simulation of these effects by platform
motion increases the cost of the motion platform and may compromise its
performance, since a higher bandwidth will allow noise in the servomechanism to produce uncommanded motions. A servomechanism moving only the seat and instrument panel may effectively provide the
simulation of these vibration effects over a frequency range of about 3–40
Hz.

6.3.4   Motion platforms for training simulators

While the requirement for six-degree-of-freedom motion in
training simulators is not experimentally established, the six-post or
synergistic motion system has been adopted for most air carrier training
simulators, and its equivalent performance required by some regulatory
authorities. Fig. 6.7 shows this system where the motion platform, which
carries the simulator cockpit, is supported at each of three points by
a pair of hydraulic cylinders. These pairs of cylinders are in turn
attached to the floor at three points, each cylinder of a pair being
attached either plus or minus approximately 60 degrees away from its
attachment at the other end (Baret 1978; Lacroix 1979; O'Dierna 1970).
Proper commanding of the extension or retraction of the cylinders can
produce both translation of the motion platform along three mutually
perpendicular axes and rotations about these axes. Motion platforms of
this geometry are made by several different manufacturers with different
cylinder diameters and lengths, so that a design appropriate to the cockpit
weight and platform performance requirement can be chosen from among
several standard designs.

The piston strokes of standard models vary from about 1–2 m which
result in maximum translational excursions of 2.7 m and rotational

excursions of up to ± 30 degrees. Note however, that each motion requires the movement of all six cylinders, so that these maxima are not obtainable simultaneously. Generally, the maximum excursions used in six-degree-of-freedom operation are limited to about 60 percent of the single-degree-of-freedom maxima. With a large enough computer capacity the available excursion in any degree of freedom can be predicted and thus maximized (Parrish, Dieudonne & Martin 1973). The servo performance of these systems can be characterized as second-order systems with a natural frequency of about 2 Hz.

The six-post or synergistic motion system is used in a number of aircraft research and development laboratories. Platform drive logic considered standard in training use can be altered to use the excursion of the several pistons to emphasize particular motions important to a given aircraft or test program. Since platform position is difficult to measure, the servomechanism position loops are closed about jack extension. However, inverse coordinate transformation can be used to calculate platform position and thereby improve performance (Dieudonne, Parrish & Bardusch 1972).

6.3.5   *Motion platforms for engineering simulators*

Because engineering simulators are used in man-in-the-loop trade-off studies of aircraft and system design, motion systems with greater velocities and excursions make possible a reproduction of a wider range of

Fig. 6.7. Six-post synergistic motion system (from Kron 1975).

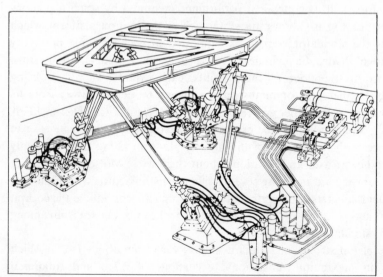

the motion cues of aircraft flight, and they are sometimes justified (Dusterberry & White 1979). Fig 6.8 indicates the frequency and velocity ranges in sway obtainable from one engineering simulator compared to those of a motion platform normally used for training simulators. The wider range of cues allows more confident and less conservative decisions to be made in setting minimum aircraft hardware design requirements. The LAS/WAVS simulator at Northrop Aircraft and the LAMAR simulator at US Air Force Flight Dynamics Laboratory use a long beam to give 6 m of an arcuate approximation of heave (vertical) and sway (side-to-side) motion of the cockpit (Gallagher 1970; Mills 1967). The Flight Simulator for Advanced Aircraft at NASA Ames Research Centre provides about 28 m in sway (Zuccaro 1970), and their Vertical Motion Simulator provides 18 m in heave. The Advanced Flight Simulator at the Royal Aircraft Establishment, Bedford, has 10 m in heave and 8 in sway.

Particular attention was paid to absence of acceleration noise in the design of the Moving Base Flight Simulator at the Netherlands National Lucht-en Ruimtevaartlaboratorium. While its excursions are smaller than some engineering simulators, acceleration noise has been held to 0.1 m/s² in heave and 0.005 rad/s² in the three rotational degrees of freedom (Koevermans & Jansen 1976).

The largest motion platform excursions and the potential for producing the best duplication of motion cues lie in the variable stability aircraft or

Fig. 6.8. Motion cues in sway from training simulator platform and from engineering simulator.

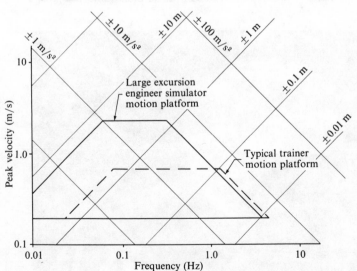

Fig. 6.9. Engineering simulator motion platforms:
(a) RAE, Bedford, Advanced Flight Simulator,
(b) Northrop Large-Amplitude Simulator – Wide-Angle Visual,
(c) NASA Ames Flight Simulator for Advanced Aircraft,
(d) Arvin/Calspan Total In-flight Simulator (TIFS),
(e) National Aerospace Laboratory of the Netherlands research flight
simulator.

(a)

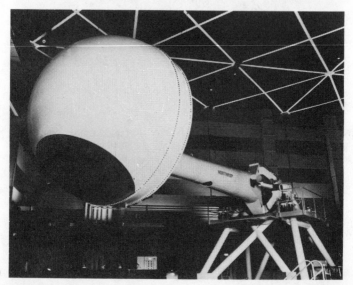

(b)

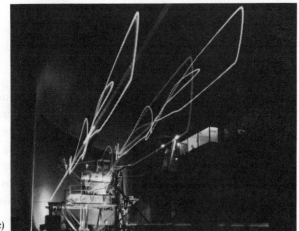

(c)

(d)

(e)

inflight simulator. This motion platform is a fly-by-wire aircraft with a computer inserted between the pilot's control devices and the aircraft control surfaces. The computer is then programmed to make the in-flight simulator's motion duplicate the motion of the aircraft to be simulated (Reynolds, Schelhorn & Wasserman 1973). Out-the-window visual cues can correspond exactly to the real world, and the pilot's display devices and the dynamics of his control devices can be made to match those of the simulated aircraft. Velocities and accelerations are basically limited by the control power of the host aircraft. The virtually unlimited excursions of this 'motion platform,' together with a perfect 'visual system,' make it a valuable tool which seems to have been particularly useful in its sensitivity to uncovering pilot induced oscillation (PIO) problems (Markham 1984; Reynolds 1982). Fig. 6.9 shows a montage of the above simulators.

6.4   Motion cue requirements

The requirement for and method of producing the inertial cues of motion in a manned flight simulator will or course depend on the projected use of the simulator. The sensitivities of the body's sensors of inertial motion peak in the range of 0.1–1.5 Hz, so the tasks to be simulated should be investigated to determine if they occur within this frequency range. Gundry (1976) and Caro (1977, 1979) have made the distinction between disturbance cues, those arising from systems failures or external environmental effects; and manoeuvre cues, those that result from the pilot controlling his aircraft. Many failures require prompt attention, and cues of these failures have a relatively high frequency content. The pilot first receives information of these failures from his tactile and proprioceptive sensors which lead the information he receives from his visual sensors. A simulator which provides him with these cues, can make him respond more realistically, and the simulator can be a more effective training device for having provided these cues (DeBerg, McFarland & Showalter 1976; Lee 1971). In stable, conventional aircraft, manoeuvre cues are of relatively low frequency, and these cues can generally be provided visually (Gray & Fuller 1977). However, in unstable or marginally stable aircraft, prompt attention is required by the pilot to maintain control, and the lead provided by the inertial cues will be important to his maintaining that control in the simulator (Caro 1977, 1979).

While there is a large body of experience on the effectiveness of motion cues in training and data indicating how motion cues affect the manner in which a pilot controls his aircraft (AGARD 1980; Caro 1977; Huddleston & Rolfe 1971; Levison & Junker 1978; Puig, Harris & Richard 1978), there are almost no results from transfer-of-training experiments showing that simulator motion is an effective training aid for most flight conditions. The difficulties of designing an effective transfer-of-training experiment and the expense of carrying one out make it probable that there will never be enough data from such experiments to serve as the sole basis for a decision on providing the inertial cues of motion in a training simulator (AGARD 1980).

In the design of a training programme, attention should be paid to those training situations where motion cues are most likely to be effective and to whether platform motion or a 'g-seat' can provide them most economically (Cardullo & Kosut 1979; Lee 1971; McKissick *et al.* 1983). Since most training simulators will simulate stable, fixed-wing aircraft, these situations are most likely to occur in the failure of automatic systems and in encounters with turbulence (Young 1978). In deciding on the simulator motion cue requirement, special attention should be paid to plans for training for emergency conditions which are hazardous to practise in flight and where the pilot could be more safely exposed to the condition for the first time in a simulator (DeBerg *et al.* 1976). The availability of data and experience in engineering simulators, particularly when they have been used for the development of the aircraft for which a trainer is being designed, should not be overlooked in gathering information on motion cue requirements (Dusterberry 1974; Jones 1979).

If some aspects of training require platform motion, the six-legged or synergistic system may well be chosen, because its ready availability can make it more economical than a system with less capability, or because the added realism of six-degree-of-freedom motion may increase pilot acceptance of the simulator. While platform motion may not be required for training in the control of the aircraft, it can increase pilot acceptance of the training device by providing cues such as landing gear extension and locking, touchdown bump, speed brake extension, and runway rumble (AGARD 1980). The strong anecdotal evidence and pilot acceptance of *good* platform motion cues in training have led some regulatory authorities to require platform motion (FAA 1980).

Many engineering simulators can make effective use of platform motion. They are used to simulate a variety of aircraft, many of which are so early in the concept or design stage that they exist only as mathematical models. At this model stage, the simulators are used in manned trade-off studies of basic aircraft stability and stability augmentation system requirements. In some conditions, the studies are of unstable or marginally stable aircraft, one of the conditions where simulator motion has been shown to be most effective. The choice of pilot control devices and instrumentation affect the ability of the pilot to anticipate and recover from failures, so design decisions are affected by the results of testing of this kind. Aircraft design decisions may be more conservative if the simulations are conducted without the inertial cues of motion. Therefore the cost savings in a production run of the aircraft may justify the cost of a more expensive motion system. However, motion simulation is only one of the tools available in making system choices, and many successful aeroplanes have been designed, built, and flown without data from such simulations. Such simulation may therefore be of most value in the design of aircraft with few applicable precedents.

While it is important to examine the potential usefulness of motion cues in simulator design, the final choice of equipment will involve many trade-off decisions. Factors to be considered in such decisions are the costs of producing the cues in a simulator and their contribution to training or research, and the cost and availability of other techniques such as aircraft flight. The large body of literature which exists will be useful in establishing motion system requirements (AGARD 1980; Puig *et al.* 1978). Staples (1978) has summarized the usefulness of simulator motions for various aircraft requirements as follows:

surge : of little use except in VTOL aircraft

sway : high frequency side force and elimination of spurious side force in roll

heave : unless a large excursion is used consider substitution of 'g-seat' or other inside-the-cockpit devices

roll : useful in tracking tasks

pitch : useful in tracking tasks

yaw : high frequency dutch roll, VTOL aircraft control, and system failures

A study of motion system characteristics which have been previously successful in the chosen application iterated with the feasibility and cost of projected motion system designs will determine a tentative motion system choice. The weight and geometry of outside-the-cockpit visual display devices should not be overlooked in the motion system design. The above study and preliminary design procedures must be carried out, even if only to limit the expense and judge the reasonableness of results of the design and validation techniques described below.

One method of estimating and validating motion system requirements, which has been carried out by Sinacori (1977), involves the use of some simulation hardware. While this is a disadvantage, few who set requirements will not have access to some level of such hardware that could be usefully employed. A simulation typical of the projected use can be carried out using only the visual cues of motion. A group of pilots can then fly the simulation task. The experiment design should include the aircraft type involved, the failures anticipated, etc. Time histories are recorded of simulated aircraft excursions, velocities and accelerations. These time histories may then be used as forcing functions in the computer simulation of the hardware and software of a proposed motion platform, and the required excursions, velocities and accelerations of the motion platform determined. The advantages of this method are that it allows the proposed design to be evaluated over a range of experiment designs, pilot skill groups, vehicle types, etc., to validate the requirement. To the extent that the motion cues aid in pilot control, this method should over-specify the motion platform. To avoid this it may be desirable to add an alternative cue of acceleration, visual, tactile, or audible, to aid the simulator pilot in keeping vehicle accelerations, and thus motion system requirements, within reasonable bounds.

Baron has used another method which extends simulation – it simulates the simulator *and* the pilot. The pilot, his visual and motion perception, control strategy and motor response are all modelled, as is the proposed simulation hardware and software. A representative flight task, such as a tracking task, is presented to the mathematical pilot, and the accuracy with which the modelled pilot can perform the task recorded in a computer simulation. Time histories of the performance of the modelled pilot are recorded and can be compared to known flight performance. Since all of the hardware is modelled, a wide range of simulator hardware can be conveniently investigated. The most economic combination of simulator hardware which produces cues meeting known pilot performance in the aircraft can then be selected. The validity of this technique lies in the accuracy of the pilot model and its applicability to the task chosen – if the model were completely accurate in all situations, it would not be necessary to have either simulators or aircraft pilots (Baron, Lancraft & Zacharias 1980; Lancraft, Zacharias & Baron 1981). The method has been used to investigate trade-off decisions in the choice of providing motion cues by visual or proprioceptive methods (Baron, Muralidharan & Kleinman 1980; Baron 1983a) and the choice between platform and 'g-seat' methods of providing proprioceptive cues (Baron 1983b).

7

Visual systems

7.1 Introduction

An essential part of flight simulation is the generation and display to the pilot in the simulator of a simulated perspective view of the outside world. The visual system which accomplishes this receives inputs from the computer which is computing the position and attitude of the aircraft as it moves along the simulated flight path and must continuously provide the view appropriate to each position and attitude.

Most of the flight simulators in the world are used for training pilots and other crew members; the rest are used for research into such topics as pilot performance under different conditions of visual imagery display, motion simulation and aerodynamic modelling fidelity. Training simulators fall into two distinct groups: civil and military. The visual system requirements are different for these two groups in that the main manoeuvre to be simulated visually for the airlines is takeoff and landing, and although this is required for military planes also, many other manoeuvres must be catered for in military visual simulation.

All visual systems have (a) some kind of store, or data base, representing the terrain and/or objects to be simulated, (b) a means for manipulating the data to obtain a correct perspective view, updated in real-time, and (c) hardware for displaying the view generated.

In the twenty-five years or so in which visual systems for flight simulation have existed, not only has development been rapid but whole simulation technologies have waxed and waned. Film systems – in which a 70 mm colour film was displayed to the pilot using servo-driven distortion optics to simulate a range of flight paths around the one used to make the film – were sold for a time in the late 1960s but were not ultimately successful, partly due to the largely pre-programmed nature of the scene displayed and partly due to the cost of making the films and replacing them when worn.

Closed circuit television (CCTV) visual systems – in which a camera fitted with a special optical system moved over a physical terrain model and the view was projected on a screen by a television projector – were developed in the mid 1960s and about 120 systems were in use by the airlines world-wide by the mid 1970s. In the late 1970s these systems were progressively phased out and now few remain. The undesirable features of these systems were the large size of the terrain model board (typically 12 m long and 4.6 m high), needing a high building, the running cost of the lighting and air conditioning required and the mechanical maintenance needed on the multiplicity of servomechanisms, the television camera and other components. Also, changing the model was not very practicable. Physical models do have certain advantages and we shall return to this subject briefly later, when considering military visual simulation.

The technology which took over from CCTV was computer generated imagery (CGI) – sometimes referred to as computer image generation (CIG) – and over 300 CGI visual systems were in use by the airlines by the end of the 1970s. The world total of CGI visual systems, civil and military, is now almost 500.

CGI can be regarded as having been born, as far as the airlines are concerned – a few military and experimental systems had been built earlier – in 1972 when the US Federal Aviation Administration (FAA) certified the first CGI visual system built by McDonnell Douglas for training use by Pacific Southwest Airlines in San Diego, California. A night scene only, showing the runway lighting pattern, was generated and displayed on a crt viewed through an optical system (to be described later) to present the image as a virtual image nominally at infinite distance. We shall return to this system (which set the pattern for future airline visual system development), after dealing with the psychophysical basis of visual perception.

7.2 The psychophysics of visual perception

This topic is concerned with the relationship between the physical light stimuli to which an observer (particularly a pilot in a simulator) is exposed and the sensations produced. Because people vary in their response, an average response or a standard observer is implied. Only an outline of the essential facts is given here; useful references are (Webb 1964) and (Farrell & Booth 1984).

Table 7.1 gives characteristic luminance values of objects encountered by human eyes. Physical limitations and cost limit the highlight luminance of simulation visual displays to a range of $1-100 \, \text{cd/m}^2$. The response of the eye with light wavelength (for luminance above $6 \times 10^{-2} \, \text{cd/m}^2$) starts at 400 nm (red) and extends to 700 nm (blue) with a peak at 559 nm (green) and,

Table 7.1. *Typical luminance values*

	cd/m^2
Sun (unobscured)	14×10^8
Average sky on clear day	6×10^3
Full moon	25×10^2
Average sky on cloudy day	16×10^2
White paper in good reading light	1×10^2
Television screen	1×10^2
Cinema screen	1.5×10^1
Lower limit of useful colour vision	6×10^{-2}
Upper limit for night vision	3×10^{-3}
Lower limit for night vision (dark adapted)	3×10^{-5}

as is well known, three primary coloured lights can be added together to make white light or any other colour within the limits set by the choice of primaries. Colour displays such as the tube used in colour television receivers and many simulation displays use this three-colour principle.

Fig. 7.1 shows the field of view (FOV) of a pair of eyes looking straight ahead. The centre of the diagram represents the centre of vision of each eye, or fovea, and the grey portions are the regions seen by the left eye only (on the left) and the right eye only (on the right). The central white area

Fig. 7.1. Field of view of a pair of eyes.

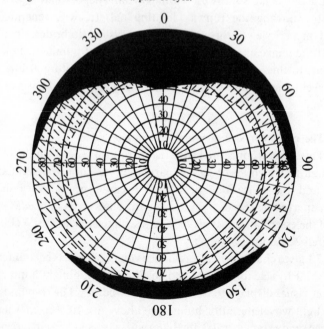

represents the region seen by both eyes within which binocular (stereo-scopic) vision is possible. The cut off by the brows, cheeks and nose is shown by the black area.

Fig. 7.2 gives the range and velocity of head and eye movements for various tasks. If the data in Figs. 7.1 and 7.2 is combined, the magnitude of the problem facing a visual system designer can be appreciated: in the real world, using movements of head and eyes, an observer can look rapidly in any direction and each glance covers a very wide FOV. The FOV sufficient for training in a simulator is very task dependent and will be dealt with later.

The next topic of importance is the perception of depth, or distances measured along the line of sight to the object viewed. The main cue to the appreciation of depth for small ranges is the disparity between the two retinal images: if one object is imaged at corresponding positions on the retinas of the two eyes, a second object at a different distance will result in corresponding parts of the retinal images having a relative displacement or disparity (measured as subtended angle at the eyes) and this is interpreted by the human visual system as a distance. This form of perception is known as stereopsis. Cues to depth due to eye muscle sensation due to focusing (accommodation) and convergence of the object fixated upon are relatively negligible (Gibson 1950).

But in addition to this binocular cue to depth, there are of course important monocular cues (one-eyed pilots exist). The main ones are:

Fig. 7.2. Range and velocity of head and eye movements.

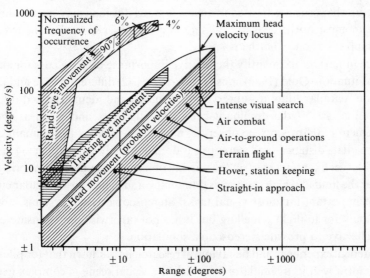

(1)   A first object that partially covers a second object must be nearer to the observer.

(2)   The apparent size of an object whose real size is known is an indication of its distance.

(3)   If one of the two objects of the same size appears smaller, it is further away (differential size).

(4)   Linear perspective (convergence of parallel edges) shows which part of an extended object is nearer.

(5)   If an object seems to move in relation to distant objects when the observer's head is moved from side to side, it must be nearer in proportion to the relative motion (motion parallax). If the observer is moving (as in an aircraft), motion parallax may be defined as the differential angular velocity between the lines of sight from the observer to the two objects, one of which is fixated by the observer.

(6)   Objects high in the field of view are generally perceived as being further away than objects low in the FOV.

(7)   Distant objects often appear bluish and hazy (aerial perspective).

(8)   A textured ground surface (e.g. grass, dense forest from above, desert with scattered cactus bushes) is perceived in its correct orientation to the observer by the differential apparent size of the texture elements.

In any situation in which an observer, stationary or moving, is viewing the real world, these binocular and monocular cues will make varying contributions to the observer's perception of where all the objects that make up the scene are located in space. Optical illusions may occur when some or most of the cues are missing and the observer makes incorrect assumptions about what he is seeing.

It is important to quantify the relative importance of these cues for any given situation. Gold (1972) chose (1) stereopsis, (2) differential size and (3) motion parallax due to observer movement and demonstrated that differential size is the dominant cue at far distances and given a slow movement of 0.5 m/s by the observer, movement parallax is dominant at intermediate distances and stereopsis is dominant only out to 17 m (64 m if the observer fixates his eyes on the moving object). Faster movement brings nearer the limit at which stereopsis dominates, and in fact the binocular cue is not important for most visual tasks. Stereoscopic visual systems have been built for in-flight refuelling but it has not otherwise been considered cost effective to provide stereoscopic simulation.

Practical experience and behavioural research have shown that for pilots to perform well in a simulator in which the visual scene is complex (e.g.

contoured terrain with multiple objects on it) it is important to provide: (a) objects of known size and (b) sufficient detail on the surface of the ground (e.g., texturing or a distribution of small objects). The objects of known size tell the pilot how high he is and the textured surface tells him the shape of the terrain (what is ahead). Recent work on this topic is reported by De Maio (De Maio 1983). High-speed low level flight by military pilots is dangerous and rapid interpretation of ground features is essential to avoid accidents. 'Dangerous' terrain includes misleading features, e.g., scattered trees that look like shrubs, so causing the pilot to fly too low.

A visually much simpler scene is an airport runway at night. An International Civil Aviation Organization (ICAO) runway and approach path is outlined by about 1000 lights, each of which appears as a point of light to the pilot. The two essential cues here are: (1) the apparent width of the near end of the runway in relation to the height of the aircraft as shown on the instrument panel and the actual width (known to the pilot) and (2) the apparent shape of the runway as seen in perspective (a trapezium). Illumination of the runway surface by landing lights from the aeroplane in the final stages of landing is also required. 'Dangerous' runways are either totally or partially slightly sloping so that the perspective shape is distorted.

The next topic relates only to a simulation display and not to the real world: it is the question of refresh and update times. Refreshing is the rescanning onto a crt of the information already written to overcome the decay of the light output due to the decay characteristic of the phosphor layer on the screen. Updating is the writing of new information.

Broadcast television systems generate the picture by scanning a number of evenly spaced horizontal lines to form a 'field' and then a second set of an equal number of lines interlaced with the first (filling the gaps) to form a 'frame'. In Europe the total number of lines in a frame is 625, the field duration is 1/50 s and the frame duration is 1/25 s. In North America the corresponding numbers are 525, 1/60 and 1/30. Since new information is written on each line, the refresh and update times are both equal to the frame time.

The reason for using interlaced scanning is to reduce flicker to an acceptable level. The eye is relatively insensitive to flicker in small areas so the effective flicker rate for the picture is the field rate; this is acceptable whereas the frame rate would give intolerable flicker. With the lower European field rate some people find flicker quite noticeable at a highlight luminance of $100 \, cd/m^2$, whereas at the North American rate no flicker is seen.

In simulator displays using a full raster (which may have 525, 625, 1023 or some other number of lines/frame) the luminance level is usually less then

$100\,\mathrm{cd/m^2}$ in which case there is no problem with either the 525 or the 625 line scanning standard; however, peripheral vision is more sensitive to flicker and this should be taken into account in designing a high luminance wide angle display.

In a visual system using CGI it is possible to halve the data throughput rate by updating at half the refresh rate (and so saving hardware cost). This may give poor movement rendition, e.g. jerkiness during rapid azimuth changes, and it is preferable to update at the refresh rate.

As will be described, many simulators do not use a full raster, but employ a calligraphic or stroke-writing type of display. Similar limits to the lowest refresh rate permissible apply as for the full raster display.

We turn next to the important factor of what is usually referred to as the resolution of the eye, in relation to simulation displays. Many different types of vision test targets are used for vision testing, each leading to a different way of characterising the ability of the human eye to distinguish detail. The preferred way to describe the minimum target size that can be seen is in terms of the visual angle it subtends at the viewer's eye in arc minutes.

Visual acuity is defined as the reciprocal of the angle subtended by the minimum size test object that can be resolved 50% of the time by an observer, and 'normal' acuity is defined as unity so that the angle resolved by a normal eye is one minute of arc. This figure may be higher or lower under various conditions.

When an observer views a simulation display what he sees depends on (a) the resolution of his eyes and (b) the resolution of the display. When the display is in raster format, it is obvious that the more scanning lines are used the greater the ability of the display to reproduce detail, up to the condition where the line pattern cannot be resolved by the observer. In practice, the ability of the display to reproduce detail is usually significantly less than the ability of the eye to resolve it and the display resolution is then the domination factor. To discuss the resolution of the display, it is necessary to describe the way in which CGI modulates the crt. The display image can be thought of as made up of picture elements or pixels, separated one from the other along a scan line by a distance equal to the scan line separation. The CGI outputs a new luminance and colour value for each pixel. The resolution of a raster display is usually defined by using patterns of alternate black and white bars or lines as the test objects and stating the resolution as the angle subtended at the viewer's eye by the image of a pair of adjacent bars or lines (one black and one white) that can just be resolved. More closely spaced bars would not be visible and more widely spaced bars would be more readily visible. (Other definitions of display resolution are used also, but the above is the one used in this Chapter.)

Considering first the (electronic) generation of a pattern of bars parallel to the scanning lines, if the bar pattern has a black bar to white bar separation equal to the scanning line separation, and the bars fall exactly on the scanning lines, the pattern will be reproduced at full contrast. If the bar pattern is moved half a line separation, it will disappear, due to each scanning line receiving half black and half white information. Intermediate positions give intermediate contrast representation of the bars. The same sort of effect occurs for a pattern of bars at right angles to the scanning lines in that there is interaction between vertical lines of pixels and the bars giving a strong or faint pattern depending on the relative position. These interaction effects are types of aliasing. Spurious effects occur for the various patterns, at all orientations, that make up a scene (Szabo 1978). Aliasing is suppressed in CGI systems by signal processing and in consequence both the vertical resolution and the horizontal resolution are somewhat less than the angle subtended at the eye by an adjacent pair of scanning lines, by a factor of about 0.7.

For calligraphic or stroke writing scanning (described later) as opposed to raster scanning, there is no fixed pattern of display pixels to interact with the information pattern and for a pair of adjacent runway lights the resolution is defined as the angle subtended at the observer's eye by the two lights when the display is just capable of showing them as separate.

Various studies under various conditions have shown that the ability of the eye to resolve fine detail falls off appreciably as the luminance level drops, and this factor must be considered in designing displays where the luminance level is low, for example in the type of display consisting of a large spherical screen surrounding the pilot where imagery is projected on the inside of the screen.

The above discussion on resolution is based on high contrast targets as presented directly to the eye or through a display. In the real world, and in simulation of it in a visual display, the contrast between two areas close to each other must be taken into account as low contrast targets have lower visibility. The usual definition of contrast is given as the ratio of the difference between the maximum and minimum luminances to the average luminance.

Using this definition, Fig. 7.3 shows how the size of a light disc target that can be detected against a large evenly-illuminated background increases as the contrast is reduced. The three curves are plotted for three values of the probability of detection, P. Such data is clearly highly relevant to visual simulation, for example in searching for a military target such as a tank against natural terrain of various types. If the target to background contrast as seen by the pilot when viewing the visual system is less than he sees in the real-world situation the training situation will be altered in that detection will only occur when the target appears larger, i.e. at shorter range. To achieve detection of a target at the correct range on a simulator visual system, the target size and/or contrast may be artificially enhanced.

For training purposes it is important to *detect* a target (see that there is anything there at all) *recognise* the target (as, say a tank rather than a tree) and *identify* it (as a particular type of enemy or friendly vehicle). Biberman (1973) reviewed studies relating the performance of a viewing system in these terms to the number of scanning lines crossing the target and to other related criteria. He quotes F. Scott *et al.* who found that 20 scanning lines gave a satisfactory level and spread of performance in identification and that there was a variation depending on vehicle type and orientation.

So far, in discussing resolution, it has been assumed that the viewer fixates on the target or on its representation on a visual system and so uses the central region of the retina, or fovea, which has the high resolution of (nominally) 1 arc minute. However, the resolution of the eye is considerably less for parts of the field of view away from the fovea, as is shown in Fig. 7.4.

Fig. 7.3. Probability of target detection.

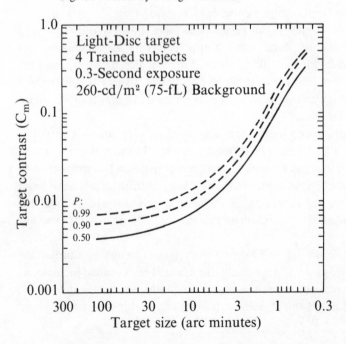

Human eyes move rapidly to fixate on different objects in turn and the perception is that the whole field of view is available at the same resolution as the fovea. The fact that the instantaneous resolution characteristic of the eye is as in Fig. 7.4 has important consequences for advanced visual systems. If the eye pointing direction can be measured continuously, in principle it is unnecessary to generate and display high detail imagery over the whole field of view, but only in the centre. This principle will be discussed later under 'area of interest' (AOI) displays.

7.3   Visual system requirements for the airlines

As has already been stated, visual systems for airline use are primarily concerned with takeoff and landing and do not need to have the capabilities (and high cost) of visual systems used for simulating the wide range of military manoeuvres. It will be convenient to discuss visual simulation for the airlines first and then to extend the description to military systems.

The image generation technique universally used is CGI. There are overwhelming advantages over other techniques: accuracy, repeatability and reliability due to the digital architecture, small equipment space occupied, minimal generation of heat, ease of changing the data base (e.g., from a representation of one airport to that of another) and reduced visibility simulation (e.g., fog).

To display the view of the outside world to the pilot, some kind of television or similar display is required and we shall deal with displays

Fig. 7.4. Distribution of visual acuity across the retina.

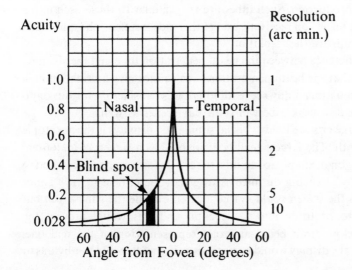

Angle from Fovea (degrees)

before discussing CGI in detail. It would be logical to work from eye characteristics as outlined in the section on the psychophysics of vision to hardware capable of presenting appropriate images. However, economic reality requires that equipment cost shall be no more then will allow for adequate training, and a practical approach is to examine the performance of available display devices and then determine to what extent these meet psychophysical requirements as related to the task to be trained.

7.3.1   *Hardware for airline visual simulation*

There are two suitable types of display device: the large crt generally of the type used for direct viewing as in a television receiver, and the television projector which projects an image on a screen for viewing. Of the large crts there are two distinct types and each has a different performance. The earliest type used, and still widely favoured in economically priced systems is the beam penetration tube which has two phosphor layers deposited on the inside of the face, one fluorescing red when struck by the electron beam and the other fluorescing green. The anode to cathode voltage is switched between two levels, the lower one resulting in the beam exciting the layer facing the cathode and the higher voltage giving penetration of the beam to the second layer. The tube can provide a limited range of colours from red, through what appears as a yellowish white, to green. This range is sufficient for training, the essential colour requirement being to differentiate between these three colours in the runway lighting pattern. Blue taxiway lights must be shown as green.

The other type of large crt used is the shadowmask type in which three electron guns excite groups of three phosphor dots on the screen to allow full colour reproduction. Such tubes are very similar to those used in the largest colour television receivers, except that closer spaced phosphor dots are used to improve the resolution.

Another difference between the beam penetration tube and the shadowmask tube is that the beam penetration tube cannot give a bright enough picture for simulating a day scene although it is satisfactory for dusk and night. The shadowmask tube will show day scenes as well.

The way large crts are used in visual simulation displays is the next topic to be dealt with. The largest readily available has a screen 26 in. (66 cm) across the diagonal. Simply to mount a tube of this size in the cockpit and to view it directly at such a distance as to give a useful field of view is not practicable as the screen could not be located outside the windscreen but would have to protrude into the cockpit and this would prevent the simulator cockpit from being an exact representation of the real one. Furthermore, the display would be unrealistic in that it would be obvious to the observer that he was viewing a flat image in front of the windscreen rather than a distant image behind it.

The arrangement first used for airline training in 1972 and now adopted almost universally overcomes these objections. This is the spherical mirror beam-splitter collimated display, shown in Fig. 7.5 (LaRussa 1964). The crt and its associated electronics are mounted with the crt face downward, the image on the face being reflected from a beam-splitter set at 45° to the pilot's line of sight, then reflected from a spherically curved mirror with centre of curvature at the pilot's eye and finally transmitted through the beam-splitter for viewing by the pilot. The beam-splitter puts the crt face effectively at the focus of the mirror so that the light leaving the mirror is collimated, i.e., all rays representing a given point on the crt face are parallel. The ideal shape for the crt face in this configuration is a spherical surface of half the radius of the mirror. The mirror may be of glass or plastic material, surface aluminised, and the beam-splitter of glass with a coating of about 50/50 reflection/transmission on the surface facing the crt and an anti-reflection coating on the other face to remove a secondary image.

The geometry of this arrangement is such that the horizontal field of view (FOV) is 48° and for the usual 3:4 image aspect ratio on the crt face the vertical field of view is 36°; however, only about 32° of vertical FOV is visible from the nominal viewing position at the centre of curvature of the mirror without vertical head movement. Scaling up the size of the crt, mirror and beam-splitter would not increase the FOV. In a typical display of the kind the mirror radius of curvature is 200 cm, resulting in 'eye relief' (distance from the eye to the beam-splitter, along the horizontal line of sight) of 120 cm; this is large enough to allow the beam-splitter, mirror and the crt with its associated electronics to be mounted outside the windscreen. The volume of space from which the pilot receives an undistorted image is such that he can move his head normally; however, a display of this type is

Fig. 7.5. Spherical mirror beam-splitter display.

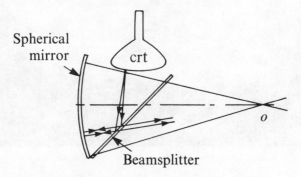

only suitable for one man, and for two pilots two identical displays are required. Fig. 7.6 shows a typical airline simulator with pilot and co-pilot displays; each has a forward and a side display.

The question of depth perception with this type of display is important. The fact that the display is collimated means that the cues that tell an observer where objects are located in space all indicate that the whole scene outside the cockpit is located at infinity: the disparity between the images in the two eyes of the pilot, his eye accommodation and convergence and the constant apparent bearing and azimuth of objects portrayed when the head is moved sideways or up and down, respectively. The effect can be very realistic. Of course, the display does not give binocular cues but, as we have seen, this can only be of consequence for flying tasks in which objects approach closely to the pilot.

Another minor feature of this type of display is that when each pilot looks straight ahead, with the plane on the runway, the identical view is seen by both pilots. The view should be slightly different due to the separation between their seats, but this small imperfection has no effect on training.

Fig. 7.6. Displays mounted on simulator cockpit.

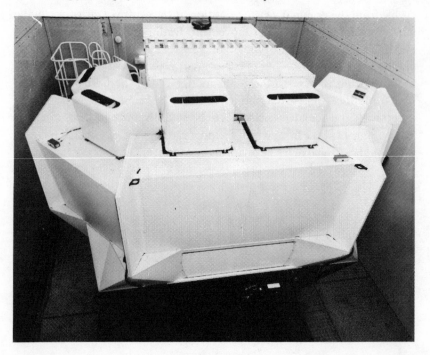

7.3.2   Computer image generation for the airlines

CGI, as used by the airlines, is a descendant of computer graphics and the display is of the calligraphic or stroke writing type as opposed to the full raster display of parallel lines used for broadcast television. However, as will be described later, the raster format is used for most military visual simulation. A review of CGI is given by Schacter (Schacter 1983).

It will be convenient to describe first how points are processed and then to extend the description to lines, polygons and polyhedra. Figure 7.7 shows a point P in the real world viewed by an observer at O, the eyepoint coordinate system being x, y, z. The point P maps to p on the viewing plane, a hypothetical plane normal to the viewing direction and distance s from it. Points A, B, C, D in the real world map to A', B', C', D' and the four-sided 'pyramid of vision' is $O\ A'B'C'D'$, the horizontal field of view being $2\tan^{-1}(A'B'/2s)$ and the vertical field of view $2\tan^{-1}(A'D'/2s)$. The display has to show all points such as p to simulate the real world. To determine the vertical distance y_p similar triangles are used so that

$$y_p = y_P s/z_P$$

where z_P is distance of P from the observer. For positions of P displaced in the x-direction a similar expression can be formed for x_P and so any position of P can be calculated, requiring, in general, two divisions. However, O represents the eye of an observer in a moving aeroplane whereas the scene viewed is defined with relation to an eye-based coordinate system. So the dimensions stated in the earth-based coordinate system, say X, Y, Z, must be transformed into the eye-based coordinate system before an image can be generated for display. This involves translation and rotation, and may also include scaling.

Translation consists of moving the centre of the earth-based coordinate system to the observer's eyepoint, requiring three additions in three-

Fig. 7.7. CGI: projection on to viewing plane.

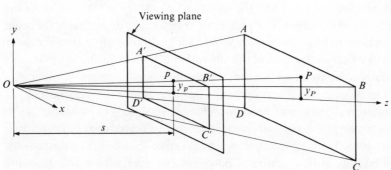

dimensional space. Rotation moves the Z-axis to coincide with the z-axis, requiring a matrix multiplication of nine scalar multiplications and six additions for every value of p.

It is usual to combine rotation and translation into one matrix multiplication using a 4×4 matrix. Homogeneous coordinates are used, i.e., a 4-element vector containing x, y and z values and a scale factor w:

$$(X \ YZ \ W) \quad \overset{\text{Rotation}}{\underset{\text{Translation}}{\begin{vmatrix} R_1 & R_2 & R_3 & O \\ R_4 & R_5 & R_6 & O \\ R_7 & R_8 & R_9 & O \\ T_x & T_y & T_z & 1 \end{vmatrix}}} \quad = (x \ y \ z \ w)$$

Thus given a data base consisting of a set of points in earth-based coordinates, the computation changes the points to the eyepoint coordinate system and then carries out the division operations to map them onto the viewing plane.

The next step is to consider the generation of straight lines. Given that the two points defining a line on the earth can be reproduced on the display, joining them together on the display produces a correct representation of the line (it is an important fact that the perspective projection of a straight line is also a straight line). A runway and approach light pattern is composed almost entirely of a group of lights arranged in straight lines or 'strings' and to reproduce a string on the display the endpoints are calculated and joined, the crt beam being moved by a display processor regularly from one end to the other while the beam current is momentarily increased at each light point position to show the light. Thus a perspective view of the runway and associated light pattern is produced, correct for the instantaneous position and attitude of the aircraft in relation to the ground, and changing in real-time in accordance with the flight path.

Projection of a line on to the display plane may result in one end of the line being off the screen. It is undesirable to allow the deflection of the beam to continue outside the display format and a clipping algorithm is usually used in which intersection of the line with the boundaries of the pyramid of vision is computed and used to clip the unwanted segment of the line.

Once a line can be mapped into the viewing plane, several can be mapped simultaneously to form a polygon and several polygons at various angles to one another in three-dimensional space can be reproduced to represent a solid object. Thus, for example, buildings can be shown around the runway, each built-up from a number of polygons. However, such buildings would be of the 'wire-frame' type used in computer graphics in which the object appears as an outline only and backward-facing polygons are visible through forward-facing ones.

First, it is necessary to remove such backward-facing polygons; this is done by computing the face normals for all polygons, computing the angle between each face normal and the line of sight and deleting those polygons for which the angle is obtuse. Secondly, occulation of parts of polygons by others without the occluded parts showing through has to be provided for as a function of distance from the eyepoint. Thirdly, the wire-frame appearance has to be overcome by filling the visible parts of polygons with scanned lines so that appropriately coloured objects can be built up. A colour has to be assigned to each polygon.

As already indicated, the first CGI systems used by the airlines showed only the runway and approach lighting pattern. However, it soon became required to include also the runway surface and markings as revealed by the aircraft landing lights during the last stage of landing. This was done by defining runway markings such as touchdown stripes and centreline as polygons and scanning the crt beam in a local raster pattern of parallel lines to fill in the polygons. Later this was extended to show a pattern of fields around the runway and to show buildings as solid objects appropriately coloured.

Fig. 7.8 is a block diagram of a typical CGI system, in which a general purpose computer receives aircraft position and attitude data from the flight simulator (or host) computer and data representing the terrain, airfields etc. over which the flight is to take place. In this part of the system the appropriate part of the visual data base is selected and passed to the geometric processor which carries out the matrix multiplication to transform the data to eyepoint coordinates and the divisions required to

Fig. 7.8. Typical CGI system.

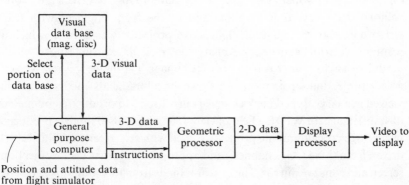

project the 3-D data onto the 2-D viewing plane. Finally the display processor generates the video signals representing the scene and the deflection signals for moving the beam of the crt display.

7.3.3   *Certification requirements for visual systems for the airlines*

In the US the Federal Aviation Administration (FAA) is the regulatory body which certifies simulators and visual systems for airline training (FAA 1983). In the UK the corresponding body is the Civil Aviation Authority (CAA) and each country has its own authority. FAA regulations regarding visual systems play a leading role in determining hardware and software requirements and some comments are appropriate here.

Certification is granted against three phases and the intention is to upgrade existing systems progressively to higher performance. The older Phase I systems required 'at least a 45° field of vision' and could use the single crt concave mirror/beam-splitter display already described. Phase II simulators must be capable of training experienced pilots so that they can transition to a more advanced aeroplane and Phase III simulators are used for novice pilots; both require a minimum of 75° horizontally and 30° vertically. Phase III displays must provide daylight scenes as well as dusk and night and must not have gaps in the field of view. The highlight brightness of the display must be not less than 20.6 cd/m (6 ft lamberts) and the resolution 3 arc minutes. To meet the daylight requirement, a shadowmask crt must be used rather than a beam penetration type and to display the larger field of view multiple displays can be used, using a modified form of the mirror/beam-splitter system. Alternatively, to obtain a large continuous horizontal field of view, television projector systems have been introduced, notably in the Rediffusion WIDE system (Fig. 7.9).

The WIDE system is a later development of the Duoview system (Spooner 1976) which was sold in the late 1970s to give a 50° wide collimated display to both pilots and to the flight engineer with television/modelboard visuals. A light valve projector was used in which a scanning electron beam deposits charge on an oil film causing ripples which modulate a light beam to give a projected image. The image was received on a back projection screen mounted above the pilots' heads and its reflection viewed in a large tipped back concave mirror so as to produce a collimated (distant) image. In WIDE a field of view 150° wide and 40° high is produced using three crt calligraphic projectors, each with red, green and blue tubes, driven from three CGI channels, and appropriately larger back projection screen and concave mirror. The mirror is made from thin aluminised plastic sheet, drawn into a frame by air suction.

The FAA requirements for the image content are as follows. Phase II capability includes all the various types of runway and approach lights including directional lights (only visible from certain directions), flashing lights such as strobes and runway end illuminator lights, rotating beacons, and vertical approach slope indicators (VASI) which show white when the aircraft is above the correct glide slope and red when it is below it. The

Fig. 7.9. Rediffusion 'WIDE' system.

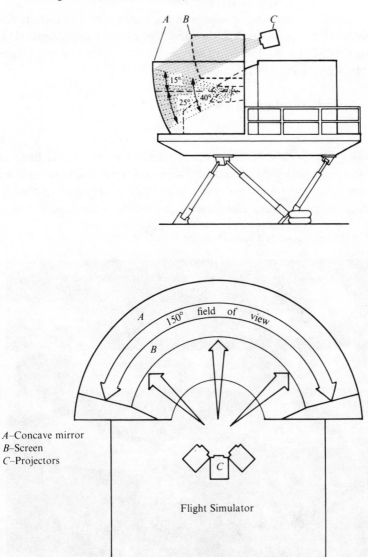

A–Concave mirror
B–Screen
C–Projectors

Flight Simulator

runway centreline lights have to be visible from 4828 m (3 miles) and the threshold lights (red and green) from 3219 m (2 miles). Plate I shows an airport scene to FAA Phase II requirements.

Phase III simulators, which have to show a day scene, have a requirement for much more detail. It must be sufficient to recognise the airport, terrain and major landmarks around the airport, comparable to detail produced by 1000 surfaces (polygons) in daylight and 4000 light points for night and dusk scenes. Typical airport scenes, generated by CGI systems suitable for Phase III, are shown in Fig. 7.10 and Plate 2.

Commercial CGI systems available at the time of writing include the Rediffusion (UK)/Evans and Sutherland Computer Corporation (US) Novoview SP3T, the McDonnell Douglas (US) Vital IV and the Singer (Link Miles Division) (UK) Image IIIT.

7.4   Military visual simulation

7.4.1   *Requirements*
As has been described, visual simulation for the airlines is primarily a matter of landing and takeoff under different conditions and can be accomplished without an extremely large FOV or very high resolution and terrain detail. Military visual simulation on the other hand covers a

Fig 7.10. Typical CGI airport scene.

Plate 1. Image II T night scene of an airport.

Plate 2. Typical CGI airport produced by image III T.

Plate 3. General Electric cell texture.

Plate 4. Honeywell computer generated synthesised imagery.

Table 7.2 *Visual simulation of military manoeuvres–typical*
requirements for fixed-wing aircraft

Manoeuvre	Terrain detail	Other aircraft or target		Total FOV
		FOV	resolution	
1 Takeoff and landing	Medium	—	—	75° H × 30° V (min) (FAA Phase III)
2 Air-to-air, high altitude	Low	15° max	High	Large
3 Air-to-air, low altitude	Medium	15° max	High	Large
4 Formation flight, high altitude	Low	80° max	Medium	80° max
5 Inflight refuelling	Low	60°–120°	Medium	60°–120°
6 Air-to-ground: weapon delivery, navigation	High	10°–30°	High	Large

wide range of manoeuvres, some very demanding, of which takeoff and landing with a fixed-wing aircraft is only one.

Military visual simulation continues to be a difficult problem, and a large commercial simulation company may typically spend half its R & D budget on visual simulation research.

Table 7.2 gives the visual requirements for a representative set of manoeuvres for fixed-wing military aeroplanes, in terms of the type of imagery to be presented to the pilot (Spooner 1979). Takeoff and landing requirements do not differ significantly from those relating to the airlines as far as the actual visual simulation is concerned, except that carrier landing is significantly more difficult and poses some special problems.

For all other manoeuvres, a useful categorization is whether an extended view of terrain is required or not. Taking first the manoeuvres which do not include terrain, air-to-air simulation at high altitude is the prime example. The pilot must have substantially unrestricted FOV of earth and sky but this can be shown with very low detail and is relatively easy to simulate. Detailed imagery (the target) is required over not more than 15°, but must be shown at high resolution (approximately 2 arc minutes). If the simulated flight path approaches the ground, however, (and air-to-air combat may well require this) a moderately detailed view of the terrain must be simulated to give reliable cues to aircraft altitude.

Formation flying training can be done without a detailed view of the ground, however, for close formation a FOV of the other plane of 80°

maximum is required, but only at medium resolution. In-flight refuelling is a similar task but needs 60°–120° FOV depending on whether a view of the whole tanker is required.

For manoeuvres requiring a detailed view of the ground, such as target acquisition followed by weapon delivery, or navigation, high detail terrain over a large FOV at high resolution is necessary. (If it is sufficient to show a single target area only at high detail, the rest of the FOV can be at lower detail). Low level flight presents the most difficult case.

Helicopter visual simulation presents a further set of demands. An attack helicopter requires a 50° downward FOV and at least 20° upward, coupled with at least 200° horizontally. For nap of the earth (NOE) flight a very high degree of terrain detail is needed, including realistic contoured terrain and trees seen from below treetop height. Other military simulation requirements include moving target vehicles which may be obscured by or may obscure contoured terrain, variable visibility including smoke clouds, hostile and friendly weapon effects and the results of target damage.

7.4.2   *Hardware for military visual simulation*

Having set out the main requirements, we shall now look at the technological solutions available and in development, with some examples of actual systems.

For air-to-air simulation, also known as air combat manoeuvring (ACM), satisfactory results have been achieved on a few trainers over a period of more than two decades by projecting an image of a target aircraft

Fig. 7.11. Air combat manoeuvring simulator.

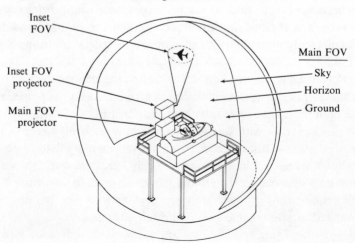

onto the inside of a spherical screen surrounding the simulator cockpit (Fig. 7.11).

This is a 'target tracked' system in which the inset FOV is placed dynamically within the main FOV available to the pilot according to the computed position of the target with respect to the pilot's own aircraft, using a servo-directed television projector. The main FOV is provided typically by a wide angle gimbaled shadowgraph projector giving a low-resolution, dim image of the horizon, sky and a suggestion of the terrain so that attitude (but not translation) cues are available. The plane image is

Fig. 7.12. British Aerospace, Warton, Twin Dome Air Combat Simulator.

derived from a television camera viewing a model aeroplane mounted on gimbals and servo-driven to take up the correct attitude. The size of the image is varied by one or more of three methods: servo-driving the aeroplane model to and from the camera, using a zoom lens on the camera and using a zoom lens on the projector. Multiple targets can be obtained with multiple target projectors and Fig. 7.12 shows such an arrangement at British Aerospace. The use of a physical model in this case rather than a CGI model was originally dictated by cost, but decreasing cost will probably lead to CGI being used for this purpose in the future.

The scene displayed is somewhat artificial in that the background luminance is typically only about $1.7\,cd/m^2$ ($0.5\,ft$ Lamberts), a luminance level at which the acuity of the eye is measurably decreased from its value at $20\,cd/m^2$ (Phase III, FAA) and so eye resolution is poor enough to be negligible in comparison with the display resolution. However, the background information displayed is of low resolution (a mottled surface representing generic terrain) so that this is of no consequence. Also the target is superimposed on the sky background and always appears brighter instead of darker as in the real world. Again, this does not affect training as the target is correctly positioned and sized and is depicted at high resolution (around 2 arc minutes can be achieved) so that its aspect can be readily seen, as is essential for training.

A spherical screen display as described shows a real image at the screen radius (typically 3–6 m) rather than a virtual image nominally at infinity, as with the spherical mirror beam-splitter display. At these values of screen radius it is of course apparent to the pilot that the image is not in the far distance but as soon as dynamic training manoeuvres commence the display becomes sufficiently realistic for useful training to be achieved.

In addition to the British Aerospace air-to-air simulator at Warton, Lancashire, England further examples are the US Navy's Device 2E6, the ACM trainer for the F-4 and F-14 aircraft, at the Naval Air Station, Oceana, Virginia, USA, which has a 350° horizontal and a 150° vertical FOV, IABG at Ottobrunn in Germany and CELAR in France.

The next step in complexity is to replace the shadowgraph projector with a second fixed television projector to give CGI imagery for the main FOV, while using a target projector as before. Where a target tracked inset is set against a main FOV the requirements are not particularly stringent, as long as the main FOV is relatively featureless, such as sky (in the case of a target aircraft), or sea (in the case of carrier landing). However, where air-to-ground tasks must be simulated, the ground target area, including the terrain immediately surrounding the target itself, is displayed as an inset at higher resolution than the main FOV and the system requirements become more stringent in three respects. First, the computation and inset projector servo requirements are more exacting as the inset image has to register with the main image during all manoeuvres, whereas air targets do not have a visual reference and small positioning errors are not seen. Second, the distortion of the image constituting the main FOV must be minimised so that when the inset takes up its correct position, the background imagery is also correct. (This requires distortion of the CGI output data usually by dividing polygon edges into segments and setting the segments at the correct angle to each other for the display to appear undistorted from the pilot's seat.) Third, the mode of insetting needs consideration because straightforward superimposition (as for air-to-air) systems may not be fully acceptable and it may be necessary to 'cut a hole' in the background image to make way for the inset.

A feature of the target inset system, of course, is that the target is shown in greater detail then the surrounding terrain. This is of no particular consequence if the approximate position of the target is known before the target is attacked but if the training calls for a visual search for the target, an unrealistic effect is produced because the target, being more clearly defined than the background, is more readily acquired than in the real world. The target inset method is thus more suitable for attacking targets than acquiring them.

The US Navy's Visual Technology Research Simulator (VTRS), located at the Naval Training Systems Center (NTSC formerly NTEC), Orlando, Florida, has demonstrated the feasibility of applying CGI to an inset display coupled with a background display on a spherical screen for military applications. Military tasks which have been demonstrated include carrier landing, formation and tactical formation flight, gunnery, air-to-ground weapon delivery and hostile environment manoeuvring. Trainers which apply this visual technology concept include the US Navy's F-18 Weapons Tactics Trainer (WTT) and the US Marine Corps AV-8B WTT. Examples of other target tracked displays include an early US Air Force low cost formation flight trainer, which presented a 90° FOV of another aircraft, and the US Northrop LASWAVS which presents a 60° FOV from a television camera viewing a modelboard.

Finally we come to low altitude simulation of high resolution, high detail terrain, the most difficult visual simulation problem. The best currently available approach to providing the required high resolution, wide (FOV) display is to divide the FOV between a number of butted displays surrounding the pilot. Each display requires its own channel of computer image generation. The number of displays and CGI channels required depends on the total displayed FOV required, the resolution required and the number of picture elements (pixels) that can be displayed in each window. Television systems with 1023 scan lines are becoming more common (as compared with the broadcast system standards of 625 and 525 lines) so that approximately one million pixels per display are available. To cover a hemisphere with imagery with resolution of 2 arc minutes per pixel (a typical requirement), approximately 24 displays and 24 CGI channels would be needed. This is not practicable on the grounds of acquisition cost alone; the system would also pose problems in setting up and maintenance which would lead to high running cost. These practical considerations restrict a multiple projector system to five to eight channels.

An early example of the multiple display, multiple CGI channel approach to wide angle military simulation is the US Air Force Advanced Simulator for Pilot Training (ASPT) which was delivered in 1974. The cockpit is surrounded by a framework of twelve equal pentagonal facets and in seven of the facets are mounted 36 in. (0.91 m) diameter crts, each viewed through collimating optics and giving a butted, continuous view to the pilot. The collimating system is known as the Pancake Window and uses a semi-reflective concave mirror and beam-splitter mounted on axis to give a flat configuration over each crt, with polarising arrangements to eliminate unwanted multiple images. The windows have only about 1% transmission and the system is limited to black and white images using the special large crts. Each crt is fed from a raster scan image generator and a total of 2500 polygon edges can be used to make the scene.

A modern example of this multi-channel method of covering a large FOV is the US Marine Corps' AV-8B (Harrier) Weapons Tactics Trainer which is capable of low altitude tactical navigation, air-to-air and air-to-ground weapons delivery, using seven light valve projectors projecting onto seven flat screens surrounding the pilot. The horizontal FOV is 240°, the vertical field 140° and the resolution in the range 12–18 arc minutes. High resolution air-to-air target projection is provided as well. The CGI used is the Evans and Sutherland CT5A raster system.

The cost of such a visual system is in the $20M range. Because of the high cost to performance ratio of this approach there has been for some years now a strong development effort by a number of organisations in the direction of area-of-interest (AOI) displays, the principle being to save cost by only generating and displaying imagery in the direction in which the pilot is looking at any given moment. These approaches are distinct from the target tracked approach which has been described above in that any part of a wide FOV can be looked at with high detail, rather than only a selected target.

The highest potential payoff is with systems that track both head and eyes as the curve of distribution of visual acuity (Fig. 7.4) can be approximated by a high resolution, high detail, small FOV of 15°–25° surrounded by a low resolution, low detail large FOV. The Singer Link system under development, with part funding by the US Air Force and US Navy, known as Eye-Slaved Projected Raster Inset (ESPRIT) (Tong & Fisher 1984), utilises a spherical screen and separate projectors for the foveal, high resolution inset image of 18° and the fixed, low resolution background image. A circular area is blanked from the background image and the foveal image inset into it.

Evaluations already carried out have demonstrated initial practicality of tracking the eye (using infrared light and a miniature television camera), tracking the head (using a radio frequency field with coil detectors on the helmet) and (a) driving the foveal projector pointing direction so as to maintain the image in the eye pointing direction and (b) commanding the CGI to generate the image to suit the instantaneous eye pointing direction. Because the foveal projector is not coincident with the pilot's head it is necessary to change the shape, focus and luminance of the inset with pointing direction. A configuration of one foveal projector, giving 1.5 arc minutes per pixel resolution (equivalent to 5 arc minutes per line pair) and three background projectors covering 270° horizontally and 130° vertically is planned. Four CGI channels will be required.

NTSC is currently developing a Helmet Mounted Display (HMD) (prime contractor, American Airlines) which has head and eye tracking arrangements and inset resolution similar to the Singer system, but projection of all the imagery is from a lens on the helmet (Breglia, Spooner & Lobb 1981; Breglia 1981). Two scanned rasters, one for the inset and one for the large surrounding FOV of 140° are generated by scanned laser beams, all the heavy components being mounted off the head and the light being conveyed to the projection system on the helmet by flexible fibre-optic guides. The advantages of projecting from the helmet are that (1) a high gain retroreflective screen can be used for returning the light to the pilot's head (so maximising light efficiency and giving high contrast), (2) there is no need to provide for dynamic change of shape, focus and luminance of the inset, (3) only two channels of CGI are required, one for the inset and one for the main FOV, (4) the main FOV moves with the head and additional projectors are not necessary, (5) due to the highly directional nature of the screen, two pilots, each using an HMD, can look at the same screen without seeing each others' images, and (6) a binocular version can be built if required. The HMD is to be integrated with the US Navy's VTRS commencing early in 1985.

Space forbids a description here of the other AOI systems under development (of which the US Air Force Human Resources Laboratory's Combat Mission Trainer system is notable), but this work has been reviewed in the literature (Spooner 1982). The AOI technology can be expected to bear fruit in the next few years and is one of the most significant developments in visual simulation.

Although CGI has mostly replaced the modelboard/television camera image generator – and CGI for military visual systems will be our next topic – some comments are in order here on modelboard/television camera systems in operation. The Royal Aircraft Establishment, Farnborough, England, has a terrain modelboard scaled at 2000 : 1 representing an actual area 20 km by 7.5 km, including individual houses and trees and allows helicopter flight down to 15 m and fixed-wing aircraft flight down to 38 m. The FOV is 48° × 36° from the camera. The very high terrain detail is valuable although the FOV is restricted.

A development carried out by Rediffusion, England, in conjunction with American Airlines for the Project Manager, Training Devices (PM TRADE) US Army, during the 1970s produced a demonstration, in 1978, of a 5000 scanning line television type moving image of a detailed model-board, presented on a curved screen 68° high and covering 180° in azimuth using scanned laser beams for generation and display (Driskell & Spooner 1976). The image was highly detailed, had a resolution of 5 arc minutes per line pair and was suitable for nap of the earth flight. PM TRADE did not continue funding this development but later funded Singer Link to carry out similar work, but to a reduced specification.

Singer Link produced a visual system for the AH-1S Cobra Helicopter Flight Weapons Simulator (Tong 1983) in which a laser beam is used to scan a modelboard over a FOV of 48° × 36° and the image obtained is displayed on a conventional spherical mirror/beam-splitter display. The advantages of the laser generator over the television camera previously used are 80% reduction in energy consumption (high intensity lamp banks are eliminated), Elimination of colour registration and image lag which were problems with the old television camera, and somewhat improved resolution from 7 to 6 arc minutes.

7.4.3   *Computer image generation for military simulation*

As described earlier, for takeoff and landing of military aeroplanes, the techniques used by the airlines are largly applicable and a proportion of the visual systems in use for military simulation used calligraphic displays driven by CGI systems of basically the same design as those used by the airlines. Landing on an aircraft carrier at night requires simulation not only of the pattern of deck lights but also the Fresnel Lens Optical Landing

System (FLOLS) which indicates the correct glideslope in a similar way to the VASI used on civilian airfields. Also the aeroplane has a hook that has to engage with one of the four wires which run transversely across the deck giving four chances of stopping the plane before the pilot has to increase power to take off and try again to land.

However, calligraphic systems cannot provide the detail required for simulating high performance tactical aircraft such as fighters and attack and scout helicopters. For these purposes a full television-type raster is required, such that a complex day scene, covering the whole screen, can be displayed. These systems are considerably more expensive than the calligraphic systems, costing several million dollars, but are justified on the high cost of $2000 or more per hour of flying military aircraft.

A basic description of the principles would follow Fig. 7.8 but instead of deflecting the crt beam to generate light point strings and locally scanned polygons, all data to be displayed (light points and polygons) is sorted out into the raster scan format before being displayed. To generate a given scan line on the display, all polygons crossed by the line must be sorted in range from the observer, so that only the nearest polygon segment is displayed, the others being rejected, for all parts of the scan line. Light points are added to the polygon segments. There are many ways of achieving the elimination of the hidden surfaces (Sutherland, Sproull & Schumacker 1974).

Raster scan visual systems available at present include the CT-5A, made by Evans and Sutherland Computer Corporation (USA)/Rediffusion (UK), the Compuscene IV made by the Simulation and Control Systems Department of General Electric (USA), the Digital Image Generator (DIG) II, made by the Flight Simulator Division of Singer Link (USA) and the Bestview, made by Hitachi (Japan) who have recently entered the market.

A description of the different architectures used in these systems is well beyond the scope of this chapter, but the overall characteristics of this type of system can be given. One of the most important parameters is the complexity of scene that can be displayed and this is expressed either as the number of polygons that can appear in the display or as the number of edges (the lines that make up the polygons); the latest systems can display about 8000 polygons at a time in one display (multichannel systems can display more).

It is important to use the polygons to best advantage, in that at any given simulated position in the data base, most of the displayed detail should be in the foreground and processing capacity should not be wasted on distant objects that are hardly visible. To achieve this, the data base is modelled in several levels of detail (LOD) e.g., a house would have doors and windows at a high LOD but would be just a block at a low LOD. The LOD used for a particular object is determined as a function of range from the pilot so that coarse representations requiring few polygons are used for distant objects. On approaching an object, the higher LOD is faded in gradually to avoid a sudden change in the appearance.

Objects built out of polygons as an approximation to the real shape have a flat, faceted appearance. To show a curved object, such as an aeroplane, Gouraud shading of the polygons is used. This is a linear interpolation process, along each polygon line segment, which varies the shading (i.e. luminance) in accordance with a chosen simulated illumination condition (e.g., diffused light from a chosen direction). Such smooth shading greatly enhances the appearance of the image.

Although a modern raster scan CGI system as described can give satisfactory training for many purposes, the imagery is undesirably stylised for low altitude flight in a tactical fighter or an attack helicopter. Instead of adding even more polygons, with the accompanying processing load, texturing patterns may be added to the polygons. Thus as well as storing digital descriptions of the colour of the polygons, 'texture maps' representing grass, concrete, sea, sky, etc., may be stored also and perspectively transformed in real-time so that they can be laid on to the polygon surfaces. General Electric, Simulation and Control Systems Department (US) call their process of this type Cell Texture.

Plate 3 shows the result obtained in a non real-time emulation; real-time operation has now been accomplished at the US Air Force Human Resources Laboratory at Williams Air Force Base, Arizona, using GE CGI equipment updated with cell texturing. A tape made frame by frame in non real-time has shown the greatly enhanced realism possible with cell texturing and the ability to show the pilot at all times where the surfaces he is viewing are located in space.

A further step beyond texturing as described is the incorporation of real-world photographic-type imagery into the scene. Honeywell, at their Systems and Research Centre (US) are developing 'Computer Generated Synthesised Imagery' (CGSI), an advanced form of CGI (Baldwin *et al.* 1983) in which photographs of real-world objects are electronically stored on video discs and the complete scene is assembled, in real-time, by (1) extracting data on all objects required for the present viewpoint, (2) sizing, positioning and perspectively warping the objects, (3) assembling the processed objects with occultation as a function of range to give the complete scene. This technique, so far demonstrated in real-time but with a limited scene content, is particularly applicable to low level flight where individual objects such as trees, vehicles and buildings must be accurately represented. Plate 4 shows the kind of result available.

A somewhat different approach to the incorporation of real-world data into computer image generation has been taken by Vought (US) (Devarajan, Hooks & McGuire 1984). Instead of using a polygon 'framework' on which to map realistic texture as with GE, or 'planting' photographic objects on a ground surface of polygons (with added texture) as with Honeywell, the Vought system uses data bases constructed entirely from aerial photographs. Work is in hand to extend this technique, which can be made to work well for high altitude visual simulation, to low altitude. It is likely that under these circumstances the Honeywell and Vought approaches will become more alike.

In concluding on the subject of computer generated images for military simulation, the topic of data bases must be addressed. For a visual system for the airlines the amount of data involved is fairly readily manageable. A plan of a runway with the lighting pattern can be used as a basis and coordinates entered into the data base using a keyboard, other objects then being added, also manually. For military simulation, the data base may have to cover a very large area of thousands of square kilometres representative of actual terrain with representative landmarks and features not required by the airlines, such as other moving vehicles.

For modelling terrain, an important source of data in the US is the Defense Mapping Agency. Terrain and culture files are available, the terrain file providing elevation information at each line intersection on a square grid and the culture file giving information about the type of terrain, e.g., city, forest, desert, water etc. To generate a CGI data base, first of all interpolation is required between the elevation locations and then the culture file has to be interpreted in conjunction with this data in a form usable by the CGI (polygons, for presently operating systems). Although DMA data is available worldwide at a scale of 250 000 to 1 and for some areas at 50 000 to 1, data at a larger scale then 25 000 to 1, needed for visual simulation, is only available for a limited number of areas.

The generation of a large military data base is a considerable undertaking, requiring several man-years. Some degree of automation has been achieved but as digital hardware gets cheaper and labour gets more expensive, the data base is becoming the most expensive component of the total visual system. There is great scope for development in this aspect of CGI.

To overcome, to some extent at least, the problem of converting DMA data into polygons, some investigations have been made into CGI architectures that operate from a regular grid data base and hence lend themselves to greater automation in data base construction. A system of this type (Spooner, Breglia & Patz 1980) has been described which has promise for further development to give high image detail of terrain. A feature of this type of system is that the various levels of detail of the data base can all be derived entirely automatically from the highest LOD by an averaging process. Also if used with an AOI type display, a very smooth transition of objects across the boundary of the AOI would be realised as the low LOD in the peripheral field of view is simply a lower resolution form of the high LOD in the AOI.

As far as hardware is concerned the whole area of computer graphics continues to develop at an explosive rate, and CGI is part of this process. Many small companies are entering the field to challenge the established companies with very much cheaper products. Very Large Scale Integration (VLSI) is reducing greatly the size and power requirements of digital hardware, to the point where complex image generation is becoming feasible during flight, for enhanced training in which synthesised targets, terrain, visual threat indication and command flight paths will be presented, with firing of simulated weapons coupled with scoring feedback to the pilot.

8

Instructor's facilities

8.1 Introduction

Instructors who instruct and assess crews in flight simulators, need specialised facilities to carry out their training duties (Bolton *et al.* 1979). These facilities and associated problems are the subject of this chapter. The chapter is based on the experiences of a particular manufacturer and similar experiences generally prevail throughout the industry.

There are broad variations in requirements between simulators: the instructor may be stationed on the simulator flight deck or off-board during the exercise; the instructor may be a flight instructor, flight engineer instructor or a tactics instructor; an exercise may take hours or it may last for only minutes. However, before considering these differences, the role of the instructor will be described in general terms so that the basic requirements can be related to the instructor's tasks and the design of the instructor's station will then be considered in detail.

8.1.1 *Role of the instructor*

Adams (1979) describes five major principles which underlie the design and use of modern high fidelity flight simulators. Two of these principles depend directly on the instructor.

The most important principle is that human learning is dependent on knowledge of results. The other principle is 'stimulus response' learning: the stimulus and the response to it are essential elements in any training exercise.

From the first principle it is essential that the instructor's facilities include means of:

(1)   monitoring the performance of the student
(2)   comparing with a performance norm

(3)   feeding back the results to the student (this may occur in the debriefing after the exercise or during the exercise itself)

The other principle of stimulus response learning is to some extent under the control of the instructor since he can change the state of the simulated environment and the state of the aircraft being simulated in an attempt to stimulate an appropriate response from the student. Means for doing this must be in the simulator.

The instructor also has the following additional tasks:

(1)   briefing the student prior to the training session

(2)   initially setting up the simulator

(3)   simulating outside sources of information for the student such as air traffic control communications

(4)   instructing and correcting the students as they proceed through an exercise

(5)   maybe filling in for missing crew members

(6)   maintaining safety standards in the simulator throughout the session

(7)   updating records

Fig. 8.1. Flight instructor's station and flight deck.

Means or aids for accomplishing these tasks are included in the instructors' facilities. In order to understand how these are accomplished the differences between instructors' monitoring techniques need to be examined.

On-board instructor

Fig. 8.1 shows the typical position of the instructor's station for a flight instructor on-board the simulator flight deck. The station is on the left-hand side. The instructor sits behind the pilots so that he can observe their movements, their instruments and in some cases the outside visual scene that is being simulated and listen to their communications. In addition he has crt displays available which present objective data on the student's progress: this data has been transformed from its 'raw' digital form in the computer to a visual form which the instructor can quickly assimilate. The control panels alongside the displays are the instructor's prime means of controlling the exercise.

Similiar comments are applicable to the other instructor who may be on board, the flight engineer instructor. In this case the instructor's observations concern the flight engineer.

Off-board instructor

The offboard flight instructor cannot see what is happening on the flight deck and he relies on displays and instruments repeating the pilot's instruments' readings, indicators showing the selected positions of important switches and some representation of the pilot's visual scene. This is all information concerning the state of the flight deck.

The instructor also needs to quickly assimilate information defining the state of progress through the exercise so that he can anticipate events and correct and advise the pilot. It follows that communication links to the flight deck are also required.

For the instructor who is concerned with tactics, the repeat displays probably include electronic displays such as radar displays and the like, since in this case it is important to monitor the student's use of the associated controls. The overall tactical situation needs to be monitored and this is usually accomplished by computer graphic displays.

8.1.2   *General requirements for the instructor's station*

From the instructor's roles outlined above it is apparent that the instructor's station is an important part of the simulator. In summary the instructor's station provides the following:

(1)   means for monitoring the progress through an exercise

(2)   means for monitoring the state of the simulated aircraft

(3)   means for monitoring the crew's actions

(4)   means for monitoring the state of the simulator

(5)   means for controlling the simulator

(6)   means for controlling the simulated environment and the aircraft configuration

(7)   means for data collection and the presentation of this data for instruction and debriefing

(8)   means for communicating with crew members

There is a human factors requirement to provide these facilities in a manageable way since the instructor is continually assimilating a wide range of important information, controlling the exercise environment and

Fig. 8.2. B747 flight simulator – flight deck layout.

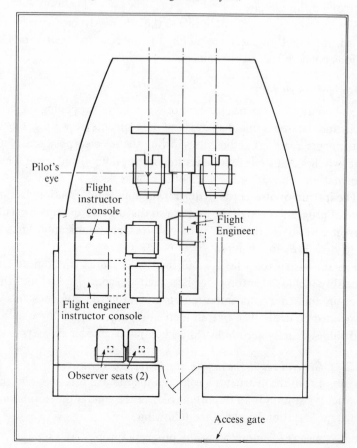

communicating with the crew. Further, the instructor's station should be unobtrusive enough not to intrude into the realism of the simulation.

8.2   Design of the on-board instructor's station

The main advantages of having an on-board instructor are that the instructor can actually observe the crew at work, the crew's instruments, the simulated outside visual scene, and feel the motion cues as the crew feels them. Thus, if the instructor's station is carefully positioned he can be in a good position to know exactly what is happening on the flight deck. Fig. 8.1 shows a typical station for the flight instructor and Fig. 8.2 shows the plan view of a simulator flight deck. As stated earlier the flight instructor sits behind the pilots where he can both observe the pilots and control the simulator environment. The flight engineer instructor (if there is one) also takes a similar position relative to the flight engineer.

The unsatisfactory element in this configuration is in the orientation of the displays and controls with respect to the instructor: whenever he uses them his attention is diverted away from the flight deck and he may miss significant information. However, there is no suitable way round this problem and the configuration shown is typical of that accepted by airlines.

In addition to the instructor's station the instructor may require a portable control unit which has a display and control capability that is adequate to control the simulator from positions remote from the instructor's station (e.g. from the co-pilot's seat).

8.2.1   Design to meet the general requirements

Let us take the requirements listed in Section 8.1.2 and examine how they are engineered for the on-board instructor's station.

Means for monitoring progress through an exercise

Typically an exercise lasts for about an hour during which time it passes through a number of phases such as initial conditions, engine start, taxi out, take-off, etc. In order to aid the instructor as the exercise proceeds, pages which describe the lesson plan and each lesson phase are available in alphanumeric form such as that shown in Fig. 8.3 which is for the engine start and taxi out phase (in this particular case the instructor is reminded to check that the external power and ground air are connected and has three suggested malfunctions listed which he may initiate and a malfunction 'armed' to act automatically). The organisation of the pages and their formats will vary between simulator manufacturers and between user airlines but alphanumeric displays are accepted in general by the users.

Fig. 8.3. Example of phase page.

```
    5    10   15   20   25   30   35   40   45   50   55   60   64

┌─────────────────────────────────────────────────────────────────────┐
│  PAGE 305 ENGINE START AND TAXI OUT                                   │
│                                                                       │
│  INSTRUCTORS CHECKS                                                   │
│                                                                       │
│     EXTERNAL POWER                              CONNECTED V1          │
│     GROUND AIR                                  CONNECTED V2          │
│                                                                       │
│  MALFUNCTIONS                                                         │
│                                                                       │
│     EXTERNAL POWER DURING ENG # 2 START                    V3        │
│     ENG # 1 HOT START                                      V4        │
│     ENG # 2 HOT START                                      V5        │
│     BRAKES OVERHEAT 1 MIN AFTER BRAKES OFF ARMED           V6        │
│                                                                       │
│                                                                       │
│     CLEAR ALL MALFUNCTIONS SET ABOVE V20                             │
│                                                                       │
│                      V21 ENGINE QUICK START                          │
│                      V23 TYPICAL MALFUNCTIONS                        │
│                      V24 NEXT LESSON PHASE                           │
└─────────────────────────────────────────────────────────────────────┘
```

Fig. 8.4. ILS approach plot.

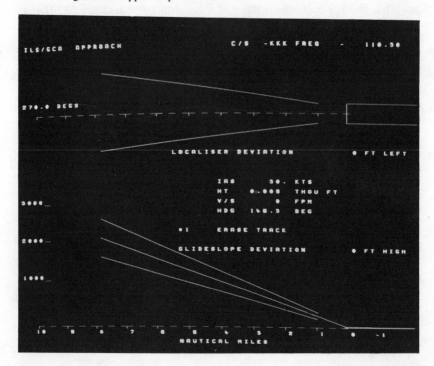

The hierarchy of the pages in the simulator is as follows:

lesson index

lesson plan

phase pages

A second display which represents graphically the situation in an exercise has also became accepted by the user. Fig. 8.4 shows such a display for an ILS approach where the instructor can see the history of the approach and quickly detect deviations in performance from a norm. More complicated and comprehensive representations are also required by various users but these are software extensions to the same display system; hence the graphics system has to be flexible enough for individual requirements to be satisfied.

In summary an alphanumeric display and a basic graphics display are required at the instructor's station and these form part of the system for much of the instructor's monitoring and control.

Means for monitoring the state of the simulated aircraft

The instructor can see the flight deck instruments and the outside visual scene and can feel the simulated motion. These provide most of the information for the instructor to asesss the state of the aircraft. However, since there are alphanumeric and graphic displays available, the information can be presented in a more precise form such as tables of parameters or in a more recognizable form such as the graphic display shown in Fig. 8.4.

Again flexibility is necessary both in the selection of parameters to be tabulated and in the form of the graphics since the requirements vary between users.

Means for monitoring the crew's actions

The on-board instructor observes the crew's actions directly and has little need of control indicators other than those already on the flight deck.

Means for controlling and monitoring the state of the simulator

Firstly we are concerned with directly controlling the simulator facilities and not with the simulation itself which is discussed later. The monitoring is straightforward via the alphanumeric display and requires little comment but the control functions do need explaining.

The two categories of control for directly controlling the state of the simulator are:

(1)   controls for the flight deck systems

(2)   controls for instructional, display and resetting facilities (using

three switch types viz., single function keys, multifunction keys, and a numerical key pad).

The first category are general controls used throughout flight simulators irrespective of manufacturer and user and includes switching on the motion system, switching on the control loading system, controlling the visual system, controlling the sound system and controlling the lighting on the flight deck. Some of these controls are classed as emergency controls and are therefore, straightforward easy-to-find switches and indicators.

The second category can be implemented in a number of ways depending on the manufacturer and user. Hence the description that follows introduces only the basic control function design and not detailed design.

Fig. 8.5. Flight instructor's console showing control panels.

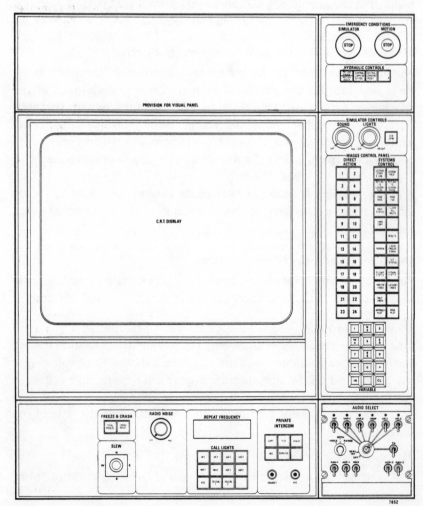

A typical layout for a control panel is shown in Fig. 8.5. The 'systems control' and 'direction actions panels' contain respectively single function keys and multifunction keys. For the latter the function depends on the page selected. Consider single function keys first. These keys are usually allocated to control functions which are used often or which are likely to be used at anytime during an exercise. For example, freezing all aircraft parameters is one of the most frequently used controls: this enables the instructor to interrupt the exercise in order to make some instructional point to the student. In this case there may either be a single key to activate the freeze function or a single key which selects the freeze page on the crt display so that the freeze may be more selective, such as selecting only height as the parameter to be frozen.

Typical controls which are selected by single function keys are as follows:

> index pages (such as the lesson index, malfunctions index, initial conditions index) usually require a single key for each index type
>
> status pages which primarily inform the instructor of the present simulated conditions may have single keys for the selection of malfunction status, flight status, visual status, route map and approach plot
>
> controls pages (such as freezes, repositioning and resets) usually require a single key for each type
>
> controls (such as master reset, auto/manual, page forward etc., clear all malfunctions, hard copy) which may be used at any time during an exercise also require single keys for each function. Similarly special pages to which a particular user wants rapid access also use single keys

Next consider multiple function keys whose functions are dependent on the page being displayed. For example, referring back to Fig. 8.3 which is the engine start page, the numbers, preceded by the symbol V refer to keys with the corresponding numbers; thus the key numbered 20 would clear all malfunctions on this page and key number 24 would advance the page to that showing the next lesson phase. Typically, up to 24 keys could be allocated in such a manner for each page.

Finally consider the insertion of variables which is usually achieved by means of a numerical key pad. The particular parameter to be varied would be selected by the multifunction keys described in the previous paragraph and the new numerical value inserted by means of the numerical key pad.

In summary the station designer needs to have four types of switches available:

(1)   emergency switches with indicators either on or positioned alongside the switches

(2)   single function keys with indicators on the switches

(3)   multifunction keys related to the particular page being displayed

(4)   a numerical key pad related to a particular line of page being displayed

The general way the switches are allocated has been described above, but individual users have their own preferences and the designer needs to take these into account. These variations will usually be implemented by software changes.

Means for controlling the simulated environment and the aircraft configuration

The simulated environment and the basic aircraft configuration would normally be determined by selecting an initial conditions page. Particular parameters on the page may be varied by using the numerical key pad. As the exercise proceeds the aircraft configuration will change as fuel is used up and malfunctions are introduced; these malfunctions may be either activated via the malfunction pages using the multifunction keys described earlier or automatically. Automatic switching is yet another technique required for the instructor's station in which the activation occurs when specific conditions are detected. The degree of complexity in automatic switching tends to be limited by the user's requirements rather than the system design: switching as a function of two or three parameters seems adequate at present.

Means for data collection and the presentation of this data for instruction and debriefing

An instructor may instruct and debrief the crew by using only the notes that were made during the exercise. On the other hand he may require more elaborate and more objective information for which the designer has to cater. The usual facilities range from hard copy devices to a software facility known as 'snapshot'.

A convenient hard copy device stores the picture directly from the instructor's display by using the display video. This is convenient since it requires no computer time, the copy is available fairly quickly and it can be initiated spontaneously by the instructor. The copy may be of a graphics plot or of a page containing performance data.

A second hard copy device is the computer printer which usually produces tabulated lists of parameters. This requires significant computer time to process and output the signal, some preparations for selecting the parameters involved and some thought as to what the tabulated output means.

The 'snapshot' facility enables the complete set of the simulator parameters to be stored instantaneously on activation by the instructor during the exercise. This enables the simulator to be reset by the instructor in order rapidly to restart the exercise at the selected position. The instructor may also record a series of frames of data which define precisely the simulator conditions during critical parts of the exercise for use during debriefing. In some circumstances the user may require record-replay facilities where the simulator parameters are recorded over a period of time (say one minute) so that the recorded period can be replayed in the simulator and the simulator flown on by the crew from the end of the replay. The replay may incorporate the movement of the primary controls and the visual. This facility is expensive, consumes computer resources, simulator resources on playback and instructor's time; therefore it tends to be offered only as an option.

Means for communicating with crew members

The instructor simulates ground services as far as the communications are concerned as well as communicating directly with the crew for instructional purposes using the internal communications. The instructor's station needs to monitor the students' communication controls (such as the VHF selected) and to incorporate a communication system so that the instructor can simulate the appropriate verbal response to the student. Amplitude control and radio noise control are also required.

8.2.2   Implementation and human factors

The design of any instructor's station will be based upon the present practice of individual companies probably with some software modification reflecting developing trends or users' special requirements. Thus there tend to be no major changes from the preceding instructors' stations. The basic station will consist of one, two, or three comprehensive display devices (such as TV monitors), standard control panels, seats for the instructors and a convenient work top. These will generally be modular so that they can be arranged in different physical configurations. The techniques for monitoring, display and control for the simulator will also be well established in the software which will be flexible enough to allow the design of pages of data and control functions to meet the user's specific requirements.

Although the basic systems vary from manufacturer to manufacturer there is much similarity between systems, which is not surprising since the same constraints and user preferences influence all the manufacturers. Figs. 8.1 and 8.2 typically show a generally accepted basic arrangement for an on-board instructor's station; i.e. the flight instructor sits behind the left-hand pilot with the instructor's controls and monitors mounted on the left-hand side. Care must be taken to prevent spurious reflections from the displays degrading the simulation.

The basic design principles which govern the detailed positioning of the controls and displays in order to provide good access to controls, good viewing and to reduce instructor fatigue and work load are given in such texts as Van Cott & Kinkade (1972), MacKay (1980) and MIL-STD-1472C. In particular, the first reference deals with human characteristics which provide the basis for man-machine designs and training device design concepts; the second examines visual factors and the third describes workspace design requirements.

The main display devices themselves are at present crt devices which are either high-quality raster monitors or cursive displays. The main advantage of the cursive display from the users' point of view is that the characters and graphics can be drawn as continuous lines which provides for a higher quality display than the dot matrix and pixel based raster system. Its main disadvantages are that it is more expensive; the range of colour is limited since beam penetration crts are used (Spencer 1981); the refresh rate depends upon the amount of displayed data; coloured areas consume computer processing time and there is a more restricted choice of suppliers compared with raster systems. For an on-board instructor's station the amount of data displayed seldom warrants a cursive system and a raster system is preferred. However, this should use high quality colour monitors so that the display can be either 512 lines non interlaced or 1024 lines interlaced. Fig. 8.4 is an example a 512 line system. Probably the non-interlaced system is adequate for most purposes but in some cases highly detailed graphics with small symbols will require 1024 lines which will usually be interlaced. Unfortunately interlaced pictures have 25 Hz intensity components which will be well above the flicker threshold (Pearson 1975, p. 45) of the instructor at even modest brightness levels (say 25 ft-l). It may be necessary to use long persistence phosphors in order to reduce the irritation of flicker to an acceptable level. It is important to realize that the instructor's view point is within 1 ft to 2 ft from the display and the sensation of flicker is worse for large viewing angles and therefore flicker critically affects the subjective quality of the display.

Also when viewed from such a close range the colour elements of the shadow mask tube (Morrell (1981) describes modern shadow mask techniques) and the raster lines are discernible and degrade the subjective quality of the display unless a small crt display is used. Unfortunately although small displays subjectively have a more pleasing image they are more difficult to read particularly if the instructor is some distance from the station, e.g. in the co-pilot's seat. Displays with at least 19 in. diagonals are probably preferred by most users. Conventional graphic display formats such as 512 × 512 pixels per picture or 1024 × 1024 pixels per picture are used with square pixels. Typically the system is capable of displaying 32 lines of 64 characters and eight colours are adequate for present usage. The minimum character font size of 5 × 7 pixels is acceptable (Gorrel 1980) but, for alphanumeric displays a better font (e.g. 5 × 9 pixels) may be preferred. Standard upper case characters are used. Gorrel (1980) suggests minimum legibility criteria of vertical angular subtense of characters of 18 arc–minute at 60 cm viewing distance, brightness 25 ft L and a contrast ratio of 4:1.

In order to improve the instructor's viewing of the display when there are extraneous light sources reflected by the display it may be necessary to place an optical filter in front of the display (Mackay 1980). This attenuates the reflected light twice and the light from the crt phosphor only once. Thus, although there is an overall loss of brightness there is an improvement in contrast; the loss in brightness may also subjectively improve the display quality since it will raise the observer's flicker frequency threshold.

A review of visual display standards is given in Rupp (1981). This brings together various standards from different authorities concerning good viewing of visual display units.

The control panels are made up of arrangements of key switches, toggle switches, rotary switches and potentiometers. Emergency switches such as 'simulator stop' and 'motion stop' shown in Fig. 8.5 are large enough to be activated quickly if the emergency should be during a lighting failure and in extreme attitudes of the motion platform.

The nomenclature on the panels shown in Fig. 8.5 will vary between manufacturers. In this particular case the keys on the 'systems control' panel are single function keys. The 'direct action' keys are multifunction keys whose functions depend on the page selected. The 'variable' keys form the numerical key pad for inserting or changing variables on the page line selected by the direct action switching.

The keys have a minimum spacing of about 1 in. and are back lit so they are easy to operate and observe in dim lighting conditions. An alternative arrangement for the multifunction keys could use a touch sensitive screen where the switch matrix is in effect a transparent overlay over the alphanumeric display. The operator touches the display face at the location of the text describing the control he wishes to activate (see Ebeling, Goldhor

& Johnson (1972) and Umbers (1977) for descriptions of suitable switching techniques).

It may help to understand the principles of the functional design of this particular display and control system if typical examples of its use are outlined.

At the start of an exercise a 'lesson index' will be displayed in which each lesson will have a number displayed; the number will define the 'direct action' key which will select that particular lesson. The instructor presses the key to select the lesson and the lesson plan will now be displayed. The lesson plan page shows the 'lesson phases' which make up the lesson and the first phase is 'initial conditions'. This is selected by appropriate 'direct action' key and when displayed the instructor can check the initial conditions and change them if need be by using a 'direct action' key and the 'variable' key pad. When satisfied with the initial conditions the instructor selects the next lesson phase which may be the 'engine start' page shown as Fig. 8.3. The instructor is now ready to run the lesson and the 'lesson start' key is pressed. The students start the exercise which may now run automatically without further action by the instructor.

However, let us assume the instructor wants to modify the exercise. From Fig. 8.3 it can be seen that the instructor can initiate malfunctions applicable to the engine start phase by pressing one of the 'direct action' keys numbered 3, 4, 5 or 23. Assuming 23 is pressed then a page of typical malfunctions associated with 'engine start' is displayed. After selecting a malfunction the instructor returns to the lesson phase by pressing the 'return to lesson phase' key and the 'engine start' page is returned to the display.

When the instructor is satisfied that the 'engine start' phase is complete, the next phase can be introduced by pressing the appropriate 'direct action' switch (number 24 in this case). So the exercise continues.

If the instructor needs a map display or the approach plot shown in Fig. 8.4 then the 'direct page select' keys named 'area plot' or 'approach plot' respectively would be used.

The display formats are different for each simulator manufacturer and tend to change slowly over the years under the normal process of trial and improvement. The mixture of page types, the page content, the logical sequence of the pages and the logical conditions for automating actions vary from user to user. Since the designer needs to satisfy each user in this repect, a flexible programming system is required. This is likely to be based upon a proprietary high level language. Typically the language provides for the following tasks to be accomplished either by the designer or the user:

(1)   cross reference access to the simulator data base

(2)   construction of the display pages and the display of data from the simulator data base

(3)   input of data to the simulator data base

(4)   linkage of pages and construction of the lesson

(5)   logical decisions based upon relationships between items in the simulator data base (the decisions may affect the display or the lesson plan or the simulation itself, e.g. it may be the automatic insertion of a malfunction)

(6)   start of map and parameter plots such as the approach plot shown as Fig. 8.4.

With reference to task (5) above, any simulated malfunction may be manual, automatic with manual override or totally automatic. An example of an automatic malfunction with override shown in Fig. 8.3 is the brakes overheat malfunction after 1 min: the malfunction will occur automatically unless the instructor presses the 'direct action' key number 6. An example of a totally automatic failure would be an instruction to simulate a wheel well fire at 600 ft; this malfunction will occur at that height if the simulator runs under the automatic mode of control.

The above tasks are usually planned and the programmes compiled at later stages in design and in the early stages of simulator use since they are dependent on the users' training plans.

The instructor sometimes has to move about on the flight deck and to control the exercise from remote positions. The portable control unit (PCU) is used in this case. The PCU is a hand-held device which consists of a switch matrix and LED display. A PCU is illustrated in Fig. 8.6.

The LED display monitors the current page number, the multifunction key number and any variable value.

The switch matrix keys typically duplicate the actions of the switches at the console: direct-page keys control the page displayed at the console, e.g. selection of flight status (FLT STAT) will cause the flight status page to be displayed; direct-function keys control the simulator, e.g. clear malfunctions (CLR MALF) clears all the malfunctions that have been activated in the exercise so far; numeric keys duplicate the action of the multifunction keys described earlier, e.g. if number 1, number 5 and the 'in' key are pressed then this activates the function associated with key 15 at the console.

The engineer instructor is mainly concerned with the performance of the flight engineer in handling the aircraft systems. In addition to the facilities discussed earlier the instructor uses schematic diagrams displayed on the console monitor as aids to monitoring malfunctions and in instructing the engineer on specific points. In general, the pilots and engineer work as a crew in the simulator and the two instructors also work together. However, in some circumstances it is convenient to separate the pilots' exercise from that of the engineer and the two exercises are independent. This independent crew training (ICT) facility is selected at the instructor's console and the two stations then work independently of each other.

8.3   Design of the off-board instructor's station

The off-board instructor has the disadvantage that he cannot directly monitor the activities in the cockpit apart from actions affecting the displays and indicators at the console. Further, he has to communicate with the crew under training using communication links instead of informal conversations as the on-board instructor does. In general, off-board

Fig. 8.6. Typical portable control unit.

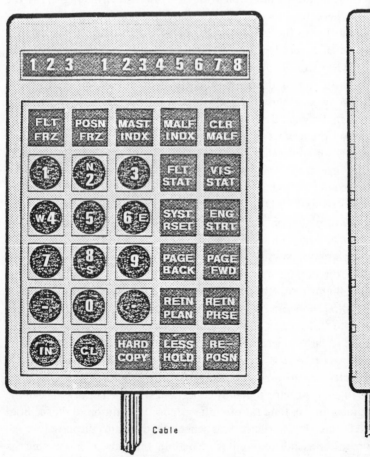

Cable

stations are confined to military simulators. It follows that the monitoring, control and communication facilities have to be more comprehensive and these areas require the designer's best talents for developing an elegant solution.

Off-board facilities may be required for radar instructors, sonar instructors, (for ASW operations) gunnery and navigation instructors (for helicopter operations) etc., as well as for the flight instructor. The discussion below is limited to the flight instructor's and radar instructor's stations since these typically include the most difficult extra facilities.

8.3.1   *Flight instructor's station*

In addition to the controls and monitors already discussed for the on-board instructor the off-board instructor typically could need the following facilities:

(1)   the monitoring of aircraft instruments, switches and controls, visual simulation, head-up display (HUD), the tactical situation and weapons

(2)   the controls for the monitoring in (1) above, for other vehicles in the exercise (e.g. opposing aircraft) and for recording facilities

(3)   the communication links to the pilot for instructional purposes and to other instructors and operators

The basic facilities generally comprise those already described for the on-board instructor's station but mounted on a larger console which is off-board in a suitable location. Typically this would be where the instructor can conveniently see the simulator so that he can quickly recognise if something untoward should happen.

Some of the additional features listed in (1) and (2) above which are likely to cause the designer difficulty are discussed further.

The monitoring of aircraft instruments requires a major decision by the user viz. should he accept a graphical representation of the instruments on a crt monitor or should he have a layout of replicated instruments arranged in a similar configuration to that in the cockpit. The former is the least expensive choice which is convenient for the designer but the latter enables the instructor to have the same view of the instruments as the pilot and it is likely that the instructor himself, being a pilot, will be familiar with the configuration.

Monitoring the position of cockpit switches and controls is generally straightforward but there are two categories of switches which it may be sensible to omit. The first includes switches or controls which only affect the cockpit equipment itself and which reflect the pilot's preferred working environment (such as the brightness controls of displays). The second includes switches and controls which cannot be monitored directly without modifying the cockpit equipment; typically these would directly link into some processing device mounted in the cockpit and it may be more convenient to monitor only the secondary effects of the control.

Monitoring the visual simulation will be costly if the full field of view of the visual is required. An acceptable compromise may display continuously at the instructor's station the channel central to the pilot's field of view and to have a second monitor switchable between the other channels. The repeat central channel may also need to be aligned accurately with a repeat HUD and in this case more problems arise. These are: the HUD will provide a collimated image while the monitor will not; the viewing through the HUD will be sensitive to the instructor's head position; since both the HUD and the monitors are analogue output devices there will be discrepancies between the pilot's visual view through the HUD and the instructor's view through the HUD.

A good compromise in this case from a human factors point of view would be to mix the signals from the HUD and the visual electronically (preferably at a digital stage) and to display the combined signal on the monitor. This would reduce discrepancies due to collimation differences, instructor's head position and differences between analogue displays. However, due to the difference between the visual display signal format and the HUD format, the mixing may be expensive to achieve and the HUD resolution may be degraded.

Another solution would be to superimpose the HUD outline and symbols on a tactical display which shows the location of important tactical features viewed from within the cockpit. This is sometimes referred to as an 'inside out' display. When used in conjunction with another display showing the tactical situation viewed from some distance outside the cockpit ('outside in' display) most of the important information required for assessing the pilot's performance is available to the instructor (Spring (1976) reports on such a configuration).

Monitoring weapon switching is done in a straightforward manner. However, although methods of automatic assessment of the pilot's performance in manoeuvring and in releasing the weapons have been under development over a long period (see Spring (1976) and Dickman (1984)), they are subject to the user's particular ideas on assessment. Therefore flexibility in the assessment software and in the debriefing display formats will be essential for good design in this area.

The control of other vehicles involved in the exercise should be limited to the selection and initialisation of preprogrammed manoeuvres if possible. However, in some exercises (such as air combat) the vehicle needs to be moved intelligently and in this case the intelligence may be provided by the instructor using simplified flight controls or by a computer (see Barnes (1984)). Again, in the latter case, flexibility in the software is needed. Recording information for debriefing is often required and this is likely to be more extensive than that required for the on-board instructor since the off-board instructor does not know exactly what is happening in the cockpit during an exercise and cannot advise immediately. The additional data may include tactical display recordings, audio records and video recordings. These may not be required regularly and may be cumbersome to implement satisfactorily at the console so their inclusion in the instructor's facilities needs to be considered carefully.

8.3.2   *Radar instructor's station*

The radar instructor typically needs additional facilities at his station to include:

(1)   the monitoring of important radar controls (such as the mode and display range selected by the radar operator), the radar display itself, the associated head-down displays, the tactical situation including the position of important radar features, electronic counter measures (ECM) sources and radar interference sources,

(2)   the controls for antenna gains, ECM sources, interference sources, IFF and extra recording devices such as radar recording systems.

The radar instructor is likely to have a heavy workload since as well as assessing the radar operator's performance he will be controlling the simulated ECM sources. Therefore, the monitoring of the radar operator's actions should be made as convenient as possible.

The most effective way of monitoring these actions is to use a repeat radar display which is likely to show in some form or other the settings of most switches and controls in the cockpit. For instance the radar mode selected will be obvious from the radar picture itself, the display range selected is also likely to be indicated together with the antenna angular position. The effects of ECM and radar interference will also be apparent on the repeat display enabling the instructor to adjust the simulated effects. More important still a repeat display will be essential for the instructor to judge the more refined ECCM activities of the radar operator such as receiver gain adjustments and manual target tracking.

There are likely to be a few important controls which cannot always be monitored via the repeat display (such as the position of the antenna stabilisation switch) and which may require extra indicators at the instructor's station. In these cases careful design will be required if modifications to the cockpit equipment are to be avoided.

The latter problem is likely to become more acute if a repeat radar display is not used and the relevant information is superimposed on a tactical display similar to that outlined for the pilot instructor. In this case special arrangements will have to be made to monitor all important switch and control positions on the cockpit equipment.

8.4   Conclusions

The facilities of an on-board station have been described. These are well established but there is scope for improvement which would enable the instructor to pay less attention to the simulator and to concentrate on the crew.

The same facilities are applicable to off-board instructors but additional facilities are required. Again there is scope for improvements which would reduce the instructor's workload such as automatic assessment, automatic vehicle control and improvements in monitoring and controlling exercises. Improvements in the design and layout of the hardware are also being developed.

The brief discussion of the radar instructor's station described more additional features which provide scope for reducing the instructor's workload. Similar comments are applicable to other specialist instructors such as sonar instructors.

New technologies and new techniques with old technologies will continue to provide the improvements.

9

Integration, testing and acceptance

9.1   Introduction

The challenge and methods used to simulate accurately the characteristics of an aircraft's flight deck and performance are systematically discussed in previous chapters. There is a further challenge involved in putting all these elements together into one package and making it function as originally intended. In terms of complexity this is analogous to putting together the aeroplane itself although for the intricacies involved in integrating an airframe, powerplant and systems is substituted the intricacies of a large computer program.

Fig. 9.1 is a diagram of simulator building blocks all of which undergo stand-alone and integrated testing before the ultimate owner is invited to survey the handiwork and pass judgement. This chapter describes how the simulator building blocks are integrated step by step into a cohesive training machine, what tests are executed along the way and the procedure used for acceptance by the customer when he evaluates the end product. The chapter is split into four sections as follows:

(1)   hardware integration

(2)   software integration

(3)   hardware/software integration and testing

(4)   simulator acceptance

This sequence of presentation follows the development and testing process of any product whether it be flight simulators or a tube of toothpaste. The level of effort required in any phase is a function of the amount of novel design in the simulator as the design and basic testing effort required for something new and unique is much greater than that of the 20th unit of a standard production item.

9.2   Hardware integration

The sequence of hardware tesing on a simulator varies depending on its construction, but basically begins with initial testing of elements on a stand-alone basis. The following analysis describes the procedure for hardware test and integration of each simulator element individually:

(a) It is apparent from Fig. 9.1 that the brains of the entire operation is the simulator computer which coordinates and schedules the operation of all other simulator hardware. The integration therefore starts by establishing a functional computer and then adding peripheral (to the computer) hardware in a logical sequence. Initial tests consist of running diagnostics, to obtain a pass/fail result on computer units such as the memory, disc controller, arithmetic unit, etc. The basic operating system is generated to match the hardware configuration, and the task scheduler is created to service on-line, background and foreground tasks. The computer is now ready to receive data files, describing the interface configuration, which are

Fig. 9.1. Elements of a full flight simulator.

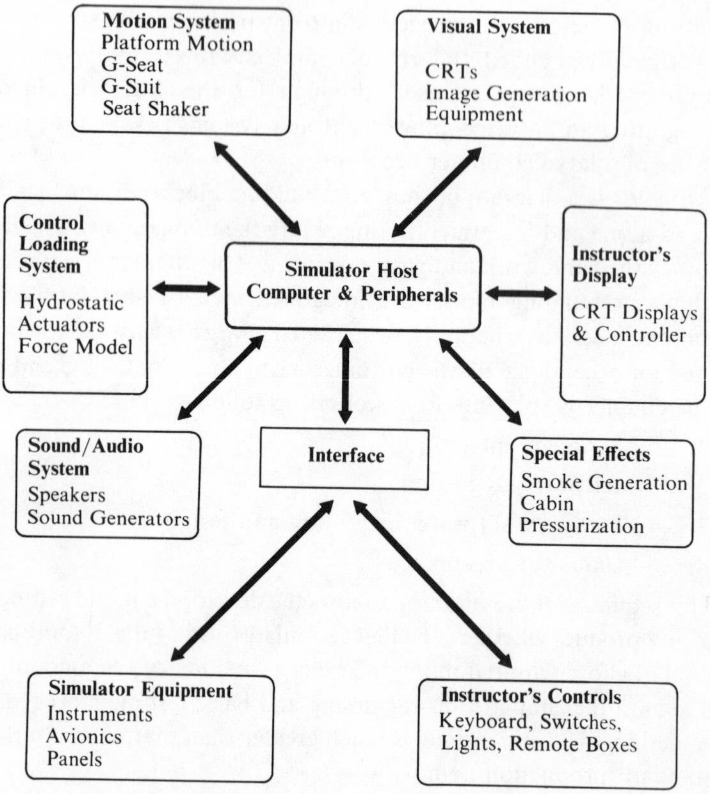

created from lists of memory locations and their corresponding interface addresses. Those files define the transfer of data between all elements of the flight simulator and the computer.

(b) The cockpit interface or linkage consists of a number of circuit boards which perform analogue to digital and digital to analogue, conversion to interface the cockpit instruments and controls to the computational system. The interface can be tested stand-alone through the use of a test set which will run through all input and output channel addresses. Diagnostic boards are embedded to perform wrap around testing by using input channels to read outputs. After stand-alone test the interface is linked to the computer which provides the addressing and read/write requests to drive the interface under real-time computer control.

Once the interface is tested and completely functional the cockpit instrumentation, switches, lights and circuit breakers are integrated with the interface. The procedure is first to all to check the wiring, then drive the instrument or light to ensure electrical integrity and then perform instrument calibration to ensure the maximum range of signal covering the instrument's full scale. The calibration is important to achieve maximum correlation between computed and displayed instrument values which would otherwise be limited by the interface resolution.

(c) The control loading system uses small stroke (typically 4 in.) hydrostatic hydraulic actuators controlled through a servo valve and linked to the flight control levers. The servo valve is positioned as a function of a mathematical model of the aircraft's feel system which may be implemented through analogue circuitry or microprocessor software. The model includes spring gradient, breakout forces and artificial q-feel systems in addition to cable characteristics, friction and the inertia of non-present aircraft parts at the control and at the surface.

Most designs of both types of model allow stand-alone testing to set up the basic feel curves and inertia terms using some form of force versus position measuring equipment in the cockpit. In the case of an entirely digital controller the testing involves the calibration of the load unit force and position transducers and establishing the null position of the servo valve. Thereafter the entire process is done via a video terminal hooked up to the digital controller which runs tuning and calibration utilities. Integration with the host computer is in the form of a digital data link transmitting parameters such as friction and dynamic pressure to the controller and receiving simulated aircraft surface actuator position in return.

(d) The instructor station panel buttons and lights are tested as part of the interface checkout. The graphics display system can also be checked out using test patterns to electrically align the crt and establish the correct operation of the symbol generator. The instructor station basic software such as crt page utilities and the pages themselves can now be installed and checked for format and colour. After interface integration the button logic, including page select and direct instructor controls, can be checked in readiness for the installation of simulation software.

(e) Motion platform systems, '*g*-suit', '*g*-seat' and seat shaker systems are all servo driven mechanisms utilizing analogue or digital servo controllers. Stand-alone tests are used to establish the nominal performance level of the equipment in terms of displacement and dynamic responses. The actuators are tested individually prior to hardware integration to reduce the amount of simulator time used on routine tuning of standard equipment. The construction of this equipment allows complete stand-alone testing with communication to the host CPU through a simple serial data path or by analogue signal. This philosophy allows addition to old generation simulators through a convenient interface and for this reason most motion systems contain embedded facilities for stand-alone tuning.

(f) The sound and audio system is used to generate aircraft sounds of frequencies up to 20 KHz and to supply station identification signals, ATIS messages, etc., to the crew members via the communication equipment and speaker system. The sophistication of sound simulation has now reached the level where a frequency analysis of recorded cockpit sounds is performed and the resulting identified dominant frequencies and bands of low frequency noise are linked with events in the simulation such as engine spool speed and level of thrust. In modern digital systems the sound is modelled as a combination of sine waves, square waves and special function waveforms to create special effects such as windshield wipers slap sounds. On propeller aircraft the phasing between engines is also of importance. Testing prior to integration is therefore used to calibrate the frequency, and amplitude of the sound generator/speaker combination which must be optimized for speaker position.

(g) The visual image generation equipment (IGE) and display system undergo stand-alone tests with the display crt or projection device separated from the display optics. The IGE is tested for capacity and scene content while the crt is adjusted for brightness, convergence and resolution. Alignment of the display optics with the pilot's nominal eyepoint is done on the simulator cockpit using a theodolite and aircraft design eyepoint data. This is aligned to give the required field-of-view after which the crt is installed and aligned. The total package is integrated with the host CPU through a high speed link across which aircraft position and attitude and environmental data is passed.

9.3 Software integration

The software element of simulation presents somewhat of an enigma to most simulator users as it is difficult to appreciate why it is so difficult to create something which does not appear as a physical structure. At least with hardware it is there and it works, or it is not and it does not. To find out if the software is error free requires a similar level of effort to a detailed engineering test of the entire aeroplane it represents.

The simulation represents aircraft systems which interact via mechanical, hydraulic, pneumatic or electrical interfaces. This software operates under the control of a simulator executive responsible for transferring cockpit information to the simulation software, running the program and then feeding the information back to the flight deck where the actor awaits the reaction. Alteration and testing of simulator software is available through utilities used for program and data file modification and which also monitor and manage the simulation software configuration. The addition of the instructor station software gives the total simulator software configuration shown in Fig. 9.2.

As the simulation is broken into systems analogous to those of the aircraft, the communication between systems is also similar in that the same information is passed. Communication is via a common data base where information can be stored by one program and read by another. This common data is also accessed by the simulator's interface which reads instrument values and converts them into the required signal type to drive the display. Values of switch and control positions are deposited in the data

Fig. 9.2. Flight simulator software components.

Software		Description
1. Operating system		peripheral handlers, file management, compilers, linkers, loaders, editors, debuggers
2. Simulator	background utilities	configuration management, radio aids stations, crt page editors
	foreground	computerized test system (CTS) real-time debugger (RTD)
3. Simulator executive		real-time scheduler
	simulation software	aero, systems, motion, visual, sound, tactics etc.
	instructors' facilities software	crt pages, panel functions, map, hardcopy

base to be used by simulation programs in mathematical and logical equations.

Much of the total software is taken from the manufacturer's basic software library and is essentially 'off the shelf '. The programs which vary from simulator to simulator and require the most testing are the simulation and instructor facilities software which change due to simulated aircraft differences and a buyer's preferred instructor interface. Integration and testing of this software takes the bulk of the pre-acceptance time on the complete simulator so it is important to ensure error free, and hardware compatible software when the programs are integrated. This has led to the development of procedures which establish performance criteria for design and testing and ensure compatible program to program interfaces.

The development of a simulator can be viewed as two parallel paths – hardware and software – which are joined together and then undergo integrated testing. Prior to this event, if some software can be joined together and run in a simulated, simulator environment, then much of the testing can be transferred to a test facility and the simulator integration time reduced. This philosophy applies to the simulator host computer simulation software excluding the software which drives equipment such as the visual or motion system and which must be present for effective assessment.

9.3.1   *Software development and pre-integration testing*

Methods of analysis, design, production and testing of simulation software have been developed by major simulator manufacturers in order to create a disciplined structure in their software. Whether this is a result of quality conscious manufacturers or customer complaints on documentation and design is purely conjecture. Military establishments in the USA insist on a disciplined approach with many checkpoints for customer approval and have created formal guidelines such as those contained in MIL-STD-1644A (1982). The procedure involves a detailed analysis of the programming requirements of each simulator element through the creation of program performance specifications. The specifications detail the simulation of an aircraft element in mathematical terms which can then be subdivided into individual software modules, each of which is individually tested. Thereafter the modules of a system – say autopilot – are integrated and tested as a complete entity and then integrated with other complete systems. The basic idea of placing the emphasis on the simulation analysis and modelling at the beginning of the design cycle is a healthy one as redesigns during integration or acceptance can prove very costly. However, the testing of an individual module such as an engine fuel controller without the rest of the engine performance modules is inefficient as the performance loops are broken into a series of programs which are individually difficult to assess. More effective testing is clearly done at the system level.

The development procedure shown in Fig. 9.3 breaks the task into several phases as follows:

Analysis phase

In this phase the simulation task is defined in terms of performance. Each element of the simulator is analysed, resulting in a mathematical design approach for aircraft systems and performance, and a detailed functional description for utilities and support facilities such as the Instructor Station software or data base compilers. At the completion of this phase the simulation is reviewed to ensure compliance with what was sold and adequate depth of simulation. In the event that data is incomplete or confusing the designer is free to propose alternative approaches to simulation for review and approval by the end user who can best assess any training restrictions which may result.

Design phase

The analysis phase defines the total simulation task in terms which can now be broken into software modules. The design phase is therefore a case of packaging the results of the analysis and establishing the means and volume of communication between those modules and the rates at which they will execute. The design therefore relates to the design of the software rather than the mathematical design of the simulation from aircraft data. The simulator acceptance test manuals (ATMs) used during software debug and testing and acceptance are developed during this phase. A description of the ATM development is to be found in Section 9.5.1.

Production phase

The modules defined in the design phase are converted into program terms through the use of flowcharts or a program design language (PDL). The format depends largely on the medium being used which could be either assembly or higher level language. New languages such as Ada are

Fig. 9.3. Software development procedure.

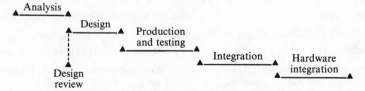

specifically designed to ensure that a PDL is used during this phase which can be easily converted into program code.

Once coded the individual modules are compiled and debugged using the test criteria established in the ATMs. After a system has been completely tested it is integrated with other simulation modules on the test facility and the interaction and resulting integrated performance is tested.

9.3.2   *Software debug and integration*

For those unfamiliar with the developer's environment during the software production phase, it consists of his design documentation, program listing, desk and computer terminal. The similarity between this and an aeroplane cockpit is only obvious to the highly imaginative, which means that other tools are necessary to stimulate the software by control inputs and assess the results normally presented on cockpit displays. A Computerised Test System (CTS) is used to establish the link between software developer and the running program. It sets up all conditions required for tests and can either create graphical plots of critical variables or perform on-line comparisons against required results and flag values out of tolerance. Development of such facilities is much simpler on a GP computer developed for the multi-user environment and adapted for simulation rather than on a computer developed for a limited application, e.g., a computer such as the DEC VAX 11/780 can support over 20 users engaged in running and testing software and retain adequate response times to terminal commands. The utilities available for this device have been created for a much wider market, including extensive university use, which means they are state-of-the-art and user friendly. Before the introduction of such computers the execution of simulator programs was limited to one or two users per computer which would run its own 'flying machine' with instruments and controls. Such an approach has severe restrictions on volume of work which modern multi-user computers have circumvented.

CTS is not restricted to the development facility but can be run on the simulator in real-time to automatically fly tests and record results specified by the test guide. This approach is used for FAA Approval test guides to automate test setups to achieve trim conditions prior to a test, to backdrive flight controls during a test and to plot out correctly scaled time histories of parameters which can be overlayed by transparencies of the corresponding aircraft flight test. Once systems have been tested on a stand-alone basis against the test guide a partial software integration can be undertaken on the software development facility under the control of CTS. This integration is designed to debug combinations of systems which have a high level of interaction, e.g., the aerodynamic model, power plant, autopilot and auto-throttle present a highly interactive performance loop which cannot be adequately assessed until they run together. Similarly the pressure profile of a hydraulic pump during an engine start is difficult to check until the power plant and hydraulics software modules run together.

One major potential pitfall with this approach occurs if the sequence of execution on the test facility does not match that of the simulator as incorrect performance can result due to the introduction or omission of phase delays. This is particularly true of the distributed process approach to simulation where critical control elements reside in separate computers. If maximum use is to be made of the test facility, the simulated environment must therefore include the executive and process structure which controls the execution of the simulation programs and interface.

9.4   Hardware/software integration and testing

Theoretically, approximately half an hour is required to successfully integrate simulator hardware and software. This is about the length of time it takes to install a disc pack on the simulator computer, load the programs and take the result for a flight around the nearest airfield. In practice the process takes somewhat longer due to the complexity of the beast and a belief that some development is best left until the simulator equipment is available. It is easier and more efficient to spot interactive deficiencies on an integrated simulator than spend time creating hardware emulations on a test facility which take longer to develop than the simulation software itself.

The integration follows a step by step build up of software which is then tested in conjunction with the simulator hardware. The process begins with the integrated hardware being driven by the common data base resident in the computer complex. The common data base is initialised to some predetermined value corresponding to some aircraft condition such as engines off, all buses powered, aircraft on-ground, ISA conditions.

9.4.1   *Environmental integration*

The first software to be added is that associated with the instructor station which enables the simulator environment to be controlled. The actions of all instructor keyboards, buttons and portable control units are tested by examining the results in the common data base, the crt display, remote display readouts and button lights. The radio aids station data base and the position update program can now be added. They define the simulated physical environment in terms of absolute position on the earth's surface and ground level altitude above sea level. The position of the aircraft and the facilities can be viewed on the instructor's CRT MAP page which depicts radio facilities type and runways using graphical symbology. Simulation of the atmosphere is added by the atmospheric program which computes temperature and pressure profiles, winds and turbulence as a result of instructor inputs and aircraft position.

In the modern flight and tactics simulator the environment has a much broader scope than that mentioned above, which is more typical of the commercial aeroplane flight simulator. An aircraft such as the Panavia Tornado is equipped with sophisticated sensors used to identify targets and threats, and perform routine navigation tasks. The simulator tactical scenario handler must command and track all the variables involved with targets including their position, radar signature, types of radar emission and ability to attack with missiles or defend using countermeasures such as flares and chaff. In simulators used for air combat training the environment includes a smart computer-driven opponent with its own flight characteristics designed to represent a known aircraft type. The opponent assesses the trainee's manoeuvre and attempts to gain the upper hand within the confines of its own flight envelope.

In all of these scenarios the trainee observes the outside world through a number of displays. These include the out-of-the-window visual displays, inertial and radio navigation equipment displays, radar, infra-red or low-light level television display, or on the display showing the picture from a camera mounted in the head of a remote controlled missile.

In the simulator the visual image, radar landmass image and inertial and radio navigation data are all computed by separate simulation elements manufactured by different companies. One of the major problems during integration is matching up those different sources of the same positional information as the equipment generating the data does not even use the same geographical coordinate system and may assume a flat earth or at best a purely spherical earth. Aircraft navigation systems are not to be fooled by this wealth of conflicting data because they know what shape the earth really is and can make corrections accordingly. The end result of this mismatch is apparent on the pilot's HUD which combines visual and mission computer data. Also the navigator's combined radar projected map display (CRPMD) combines radar returns and mission computer data overlayed on a moving map. The mission computer target position computed from radar and navigation information is depicted by a symbol on the HUD which is supposed to cover the target on the visual display when the crew has performed all the correct procedures. Similarly the cross on the navigator's display must overlay the target radar return, and the moving map and raw radar returns should coincide. The interaction is schematically shown in Fig. 9.4.

Another conflict apparent in low flying strike aircraft is the mismatch between the visual and radar landmass data bases. Radar landmass data bases use a mass of real-world information covering a gaming area over which the training missions are carried out. The level of detail is accurate to within 10 m in elevation of the real-world and cultural features such as bridges and roads are also present. In addition to ground mapping radar this data can be used to simulate a Terrain Following Radar which scans the terrain ahead of the aircraft and feeds elevation data to the Autopilot/Flight Director system. This in turn alters the aircraft's elevation profile to maintain an approximately fixed altitude above the ground. Current visual image generators are unable to match this degree of accuracy in real-time due to the update rate required of a visual system

Fig 9.4. Panavia Tornado simulator tactical display interaction.

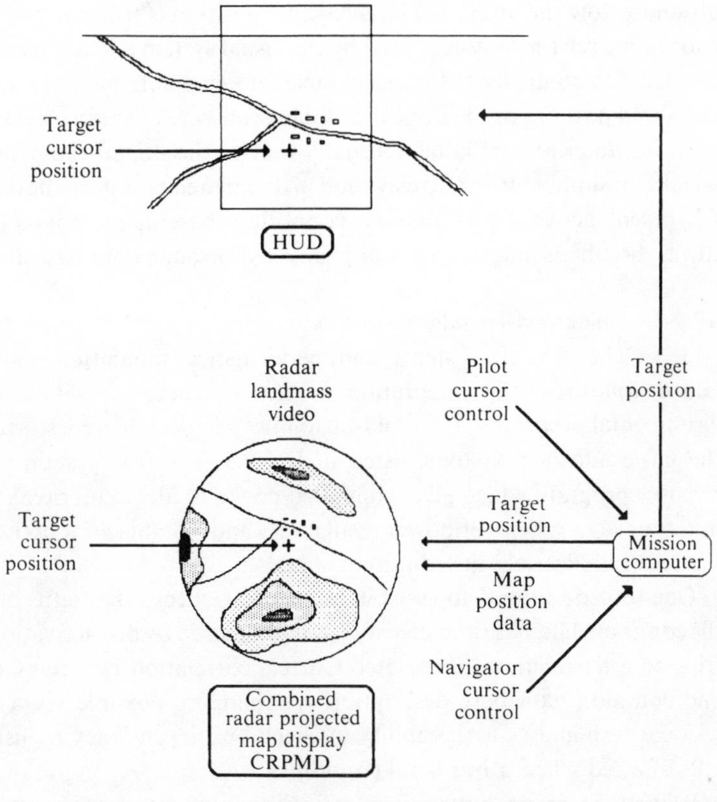

display (30 Hz) compared to that of a radar. The net effect is a ride on a roller coaster over what is apparently smooth terrain.

On the commercial aeroplane flight simulator the data base mismatch is most obvious when correlating radio navigation indications and visual positions relative to the runway touchdown point. The other most common problem is elevation where the visual system assumes a level data base and the flight simulator has an accurate profile of the runway. Unless this is taken into account the simulated aircraft can end up landing on the glideslope but visually below the airfield.

On the FRG Panavia Tornado program the synchronization of the visual, radio navigation, and radar and target data is known as harmonization.

Cohesion among data bases is achieved by selecting one source of positional information as the master to which all other data bases are slaved. The usual selection for a commercial flight simulator is the radio navigation facility data base which is used to give range, bearing and depression between a facility and the aircraft, and the elevation of the ground below the aircraft. The aeroplane position is transformed into the coordinate reference system used by the visual system and a table of offsets is created to compensate for inaccuracies at key points such as the runway threshold position or the slope of a sloping runway. The reference data base on a full tactical simulator is usually that of the digital radar landmass which can supply range and elevation of designated points on the data base. Alignment between this and the visual data base at set points (such as towns, bombing ranges, etc.) is implemented through data base alteration.

9.4.2   Simulation module integration

The aircraft systems and performance simulation software is added following the integration of the instructor facilities and environmental programs. The build-up follows a logical process starting with the basic ancillary systems listed in Fig. 9.5. The first system is in the electrics program which gives simulated power to all circuit breakers (CB) in response to generator power availability and suitable crew activation of the cockpit electrical distribution controls.

One tool, developed to assist in integration, checks the status of CBs in the common data base and displays changes caused by bus activation or CB trips on a terminal or line printer. Correct correlation between CB labels and common data base description are therefore possible via a cockpit resident terminal as is the ability to check bus dependancy by listing the CBs affected when a bus is taken off-line.

When the electrics system is integrated the simulator can be electrically powered from a ground cart. The next systems to be integrated are usually the APU, fuel and pneumatics systems which, when complete, allow the engine to be installed together with all the other ancillary systems and the secondary flight controls.

Prior to the start of Software/Hardware integration the control feel system has been tested and calibrated against aircraft design and recorded aircraft force versus position data. The calibration covers both the static and dynamic response of all the primary flight controls and nosewheel steering measured both at the control and the servo controller to ensure correlation between controller performance, measured simulator cockpit force and measured aircraft forces.

The primary flight controls program of the simulator is introduced after ancillaries and the surface position versus control position is measured against the simulator test guide for all on-ground, static conditions. Extensive use is made of CTS during this phase to produce surface and force versus control position plots. The air data computer, flight instruments and radio navigation receivers are also introduced in readiness for the

Fig. 9.5. Systems breakdown of simulation software.

4	Aero	
27	Flight controls	
22	Autopilot/Flight Director	
34	Navigation	radio
		inertial
		omega
		doppler
23	Communications	
71–81	Power Plant	
28	Ancillaries	fuel
24		electrics
49		APU
29		hydraulics
26		fire
30		anti-ice
27		secondary flight controls – flaps/slats
35		oxygen
32		landing gear and brakes
33		lighting
31	Flight Instruments	
36	Pneumatics	
21	ECS	
	Weapons	air-to-air
		air-to-ground
		fire control
		stores management
	Sensors	radar
		LLLTV
		IR

simulator's first flight and the harmonization between radio facilities and visual data base is checked on ground around the simulator's primary airports.

As systems are added the designer performs the on-ground portion of the ATM to check the performance of his system. With the philosophy of STF integration the system integration is over very quickly and much of the time is spent on initial checks by the manufacturer's evaluation pilot.

Finally the simulator is ready for the introduction of the aero package. When the aero is introduced it has already undergone extensive test facility testing with both the autopilot and control force programs which usually means the integration and first flight are anti-climactic taking only a couple of hours. The kind of problems normally encountered are erroneous scaling of flap positions or a sticky brake which wasn't obvious before as the simulator had never tried to move. Once the simulator can fly and be positioned in the air the in-flight testing of the systems integrated can be completed against the ATM.

As all the simulated components of the aircraft performance loop are now present it is possible to begin flight testing. The checks fall into two categories:

(1)   basic flight performance checks such as stabilized speeds, stabilizer settings versus power settings for different weight and c of g, stabilizer settings for takeoff, etc.

(2)   reproduction of tests run on the actual aeroplane, including environmental conditions, and checking the pilot input and aircraft response to a particular manoeuvre. The scope of this test is covered by FAA Phase II requirements (FAA 1983) and includes tests such as engine-out takeoff and cross-wind landings.

These tests plus the subjective evaluation of knowledgeable pilots (usually customer pilots) are necessary to assess the integrity of the mechanized data and the impact of man-in-the-loop control delays between pilot input and simulator response. These delays are caused by:

(1)   sampling frequency of the primary flight controls program of the controls position

(2)   execution of the control loop including primary surface computation, aero, position update and visual program transformation which may be spread over a number of processes in a distributed processor computer configuration giving an excessive throughput delay

(3)   visual image generation equipment and display which can cause between 50 and 80 ms delay

Complete subjective evaluation testing is possible after integration of the motion system software which uses flight accelerations to provide the pilot with attitude and acceleration stimuli. The phasing of this stimulus relative to visual and instrument response, and the absence of spurious motion during washout of previous motion displacements to neutral is important to avoid generating false motion cues.

9.4.3 *Advanced avionics and tactical systems*

The final systems to be integrated are the advanced avionics and tactical group which include the following:

thrust control computer (TCC)

autopilot/flight director (AFDS)

flight management computer (FMC)

radar

passive radar warning (PRW)

weapons and weapon aiming

stores management

countermeasures

Advanced avionics systems integration

The first two of these systems provide the link between positional guidance data and the control of the aircraft and so form part of a critical control loop. With the present increase in popularity for using aircraft hardware units in the simulator, the simulator must generate accurate electrical stimuli for the units as they are designed to deactivate if input errors are detected. The boxes must experience similar aircraft and engine response characteristics, including throughput delays, as the control loop gains developed for aircraft use cannot be adjusted without opening the box. The margin of stability here is low due to a build up of delays on the critical control path consisting of the following program sequence – inertial reference signal computation (software) – autopilot (hardware) – control force simulation (hardware) – flight control surface position computation (software) – aerodynamic computation (software) – inertial platform simulation (software). A difference of 20 or 30 ms in this control path can make the difference between a stable and unstable loop.

Flight management computer integration forms part of an increasing number of modern commercial simulators. This device contains flight plan and vertical profile information used to direct the autopilot. The crew can call up flight plans from a selection of airline company routes or alternatively enter routes manually.

In aircraft, as in simulators, the associated computer programs go through updates and many airlines prefer to use the same modification procedure for both aircraft and trainer which is one of the reasons for the use of aircraft equipment. Unfortunately, the computational aircraft equipment is designed to be installed and flown in an aircraft and doesn't make allowances for simulator functions such as a Flight Freeze which inhibits the integration of aerodynamic parameters such as attitude, altitude and position. Faced with unnatural situations the unit will make an assumption that either the sensor inputs are wrong or that windspeed is equal and opposite to indicated airspeed which explains why the ground speed is zero.

Two alternatives exist when an aircraft box is unhappy in the simulator – either put a training limitation on the simulator or modify the box software so that it is happy in the simulator. The choice is one which must be made by the user as only he can estimate the penalty of the training restriction.

Tactical systems integration

The tactical simulation to be integrated includes the simulated sensing devices, the performance characteristics of weapons and opponents and the interactive intelligence of simulated friend or foe. The primary sensor system is radar which scans a simulated radar landmass containing active targets. Interaction between the radar stabilization system and the aircraft attitude must be accurately simulated to produce the correct radar sweep angles. In addition the correlation of visual, landmass, navigation computer and simulator inertial position all come together on both the pilot's HUD and the navigator's radar/map display as previously described and illustrated in Fig. 9.4.

On the Panavia Tornado simulator the alignment of data bases and aircraft navigation and weapons computers is performed by placing the aircraft at a known flight position and aligning with a known visual feature (such as the ATC tower) through the HUD. The navigation computer is fed the correct aircraft position and altitude and the position and altitude of the reference point.

This data is used to drive a moving map which shows a symbol on the reference point position which verifies the correct integration of the navigation computer. This computation is also used to position a symbol on the HUD which visually overlays the same reference point in the visual scene when the visual and navigation data bases are aligned. The radar landmass data and moving map are combined in one display and therefore visually checked for correlation. The complete operation is checked by using the radar to designate the reference point and checking the position of the HUD symbol against visual and then allowing the pilot to do a visual fix and comparing the radar return against the fixpoint symbol at the navigator's position.

Weapon sensors can be TV, infra-red or radar seeking and are linked either to the weapon's own guidance control system or to a manual control. The accurate trajectory of the weapon may be necessary in one-on-one combat training where part of the exercise is to avoid hostile weapons. Accurate simulation of weapon aerodynamics and the relative positions of target and weapon are therefore necessary both for accurate weapon steering simulation and to determine the effectiveness of countermeasures such as chaff or flares.

To complete the technical scenario the automatic interaction between trainee and threat environment is introduced so that a target can react to the trainee's action. In the case of an airborne opponent the target assesses the trainee's attempts to get into a firing position and will react with countermeasures designed to place the trainee at a disadvantage. In addition the environment contains surface-to-air missile sites which use radar to seek out and track the trainee prior to missile launch and will change radar modes depending on which countermeasures the trainee selects.

ATMs are written for all the tactical systems to ensure correct system operation and interaction with the simulated world. The ultimate test comes when the integrated simulator is flown on a training flight involving navigation, malfunctions, emergency procedures and foul weather. For the tactical aircraft the final test is a mission in a hostile environment against an intelligent adversary which he can only successfully complete using all facets of his available equipment and aircraft performance.

9.5   Testing and acceptance

A typical modern flight simulator is probably tested four times before being accepted and then tested at least another once by the Approval Authority before being put into training. As the evaluation testing alone will probably cover some 19 000 test results, the question might well be asked as to why this is necessary. After all, when a customer takes delivery of a new aircraft, a 3–4 hour test flight is the norm before he signs for it. Why take an average of 250 hours just for the acceptance evaluation of a simulator representing the same aircraft? To answer these questions, we must define the objectives of the testing and acceptance.

9.5.1   *Evaluation testing and acceptance testing*

Evaluation testing and acceptance testing in-house and on-site can be treated as essentially being the same testing repeated three times, but with different emphasis in each case and approval testing is a subset of this. Evaluation testing is the testing carried out by the manufacturer's personnel prior to calling the customer to commence his acceptance of the simulator. Usually the customer 'accepts' the simulator at the manufacturer's factory ('In-house' or 'In-plant' Acceptance) and then again following its installation at the customers' training facility ('On-site' or 'Final' Acceptance). The Evaluation Test is the test period which, for the first time, treats the simulator as a full aircraft. Although primarily operational in nature, the tests also confirm compliance with the technical specification in the areas of non-aircraft systems such as the instructor's station functioning. Before any of this testing can commence, an Acceptance Test Manual (ATM) has to be written and when agreed by the customer, this is used as the basis for Evaluation Tests. This ATM generation task can begin as soon as the data analysis is completed and system design has started. Usually, the system engineer will initiate the ATM production in conjunction with the manufacturer's evaluation pilot, if he has one.

The ATMs are prepared from aircraft design and performance data plus operation manuals on normal and abnormal procedures. The format takes the form of action and result and the scope is intended to cover the testing of each simulated element through the use of cockpit equipment. This involves testing every circuit breaker, all modes of operation of equipment plus a complete performance checkout of the aerodynamics, control feel, auto-pilot stability, power plant transients and weapon and sensor performance. The ATM contains a chapter for each simulated system listed in Fig. 9.5. In addition the operation of the instructor facility is checked against the operator's guide with additional tests to check out the development facilities such as those to add and modify instructor crt pages.

The scope of aero and control feel performance tests is now more rigidly defined by certifying authorities who have introduced minimum standards of simulation required for different levels of training. These include (FAA 1983) a checklist of tests required of a simulator to give effective transition, recurring and initial training (Phases, I, II and III) without the need for any time on the aeroplane.

For the military simulator, however, the interaction between the aircraft and the environment is more easily demonstrated by flying a complete mission involving all facets of the simulation including the interaction between mission computers, simulator computers, visual displays and a simulated radar landmass. A test guide containing all mission combinations is prepared which may cover several sorties and will form the final test of simulator acceptability.

The tests in the ATM need to be arranged in a logical and consecutive order which, whenever possible, follows the operating procedures of the aircraft type and those of the customer airline/airforce. Each aircraft system or sub-system will be checked separately, but there does need to be some cross-reference to associated systems, e.g., when checking the aircraft fire protection system simulation, the pulling of a fire handle must be shown to close the fuel valve, trip off the electrical generator, close the pneumatic bleed valve, close the hydraulic pumps down as well as having armed the fire extinguishers. Typically, the sequence will begin with On-Ground checks with all simulated electrical, hydraulic, engines and pneumatic power removed – in other words, a 'dead ship'. During 'dead ship' checks, absence of simulated aircraft lighting and power off readings of instruments can be confirmed. From this point, testing proceeds through the various stages of powering up, to taxiing, takeoff, climb, cruise, descent, landing and shutdown.

9.5.2 *In-house acceptance*

In these tests, the customer, sometimes with the assistance of the manufacturer's team, carries out the tests in the ATM previously agreed as being necessary to prove compliance with the specification in all aspects. As the ATM is used for the manufacturer's evaluation testing, the customer testing covers the same areas, but the customer brings his wider operational experience (and perhaps experience on the same aircraft type), to bear. This may be significant in military simulators where current combat training/experience is not readily available to simulator manufacturers.

9.5.3 *On-site acceptance*

The customer's team now repeats a sampling of the in-house testing in order to ascertain that the move from the manufacturer's plant to the final site has not adversely affected the simulator. Sometimes on-site acceptance takes the form of running full training details or Lesson Plans in lieu of repeating ATM tests.

9.5.4 *Approvals*

Approval is usually given by an independent team from an airworthiness authority or military test establishment. After perhaps 16 hours of testing, this team indicates the type of training for which the simulator may be used. Its tests therefore are not only to compare the simulator performance with that of the aircraft, but also to examine the effectiveness of the training facilities available from the instructors' station.

Let us now look in greater detail at some aspects of the testing.

9.5.5   *Acceptance testing*

The aim of this phase must be to prove that the simulator meets the requirements of the procurement specification (which is usually an all embracing document, including such phrases as 'represents the aircraft performance and systems according to the approved data'). In addition, commercial simulators have to be approved as satisfactory training and checking devices by the regulatory authority, whilst military combat aircraft must be proved as being effective combat training devices. The first requirement, that of proving simulation of the aircraft, i.e., the Performance Validation, is conceptually the easiest. Part of the approved data is design data whilst another part is checkout data. It is the latter which is used to prove the simulation performance. Generally, this data needs to be supplemented by some subjective checking. In recent years, the requirements of the US FAA Advanced Simulation Plan, adaptations of which are being used increasingly by other authorities, have mandated objective testing which takes the form of time histories of simulator response being compared to aircraft performance recordings of the same manoeuvre.

Performance validation sequences

Each system and sub-system should be checked for:

(1)   power off conditions

(2)   power on – ground conditions

(3)   power on – flight conditions

but there is a necessary sequence of checking systems. For instance, although the atmosphere is not part of an aircraft, it is vitally important that it should be proved to be correctly modelled prior to testing any aircraft system which reacts to the atmosphere such as flight, power plant, flight instruments, pressurization, air-conditioning, etc. A suggested sequence is shown in Fig. 9.6. From this figure, it may be seen that whilst any system can be tested in a stand-alone situation, some areas are so interrelated that the only true test must be of a fully integrated nature. Thus, it is no use checking the autoflight system unless flight has been proved. Flight performance is dependent upon power plants having been proved and power plants in turn, cannot be said to have been proved unless the takeoffs occasioned by pressurization bleeds and hydraulic pumps are tested as correctly simulated.

Fig. 9.6. Aircraft systems test sequencing.

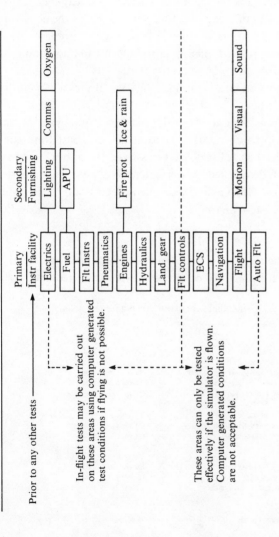

1. Primary systems sequencing is critical to efficient testing

2. Secondary systems have no interdependency but should only be tested after primary systems (to the same level) are usable.

3. System malfunctions are tested with the related system.

4. Use hardcopy/plotting devices to record results wherever possible.

As almost every system needs some data input in order to operate, it is necessary that the instructors' station is fully operational and checked prior to starting any system checks. Luckily, the instructors' station is one which may be checked in a stand-alone condition, except for a minority of areas, so it is ideally the first of all parts of the simulator to be tested. During customer acceptance, it is a good idea to combine part of the checking with a 'familiarization course' for the customer's acceptance team. The correct usage and understanding of the instructors' station speeds the consequent testing and reduces the number of times it is necessary to repeat any test results which have been proved unacceptable, not due to poor simulation, but because of incorrect setup or operator error.

The next most important task is to check the basic electrical system. This really requires all circuits to be checked by opening each CB in turn to prove bus bar feeds, transfers and system dependencies. A long and intricate task, but one which ultimately reduces the total task. It is best done using a power distribution chart as in Fig. 9.7. At this stage, the only effect to look for is that when the CB is pulled, power is lost to all systems or parts of systems fed through the CB. The more detailed effects upon those systems are checked in each system chapter.

Performance validation methodology

System checks are the simplest because one is usually dealing objectively with them – looking for pressures or temperatures associated with certain

Fig. 9.7. A300 B2K —200 flight simulator acceptance test manual.

D. Circuit Breakers

TITLE	LOCATION	CB #
AC BUS & TRANS WARNS		
AC BUS & TRANS WARNS	124 VU, D14	18XG
AC BUS TRANSF CONTRL& EXT PWR ON WARN	124 VU, D10	19XG
EXT PWR/CONTRL & AVAILABLE WARN	124 VU, D13	20XG
AUX TR CONTRL&GRD EXT PWR WARNS	124 VU, D11	21XG
115 VAC BUS 1		
AC VOLTAGE & FREQUENCY IND/BUS 1	124 VU, C19	13XV
GEN 1, AC BUS 1 WARN RELAY	124 VU, E7	29XN
115 VAC BUS 2		
AC VOLTAGE & FREQUENCY IND/BUS 2	124 VU, C20	14XV
GEN 2, AC BUS 2 WARN RELAY	124 VU, E22	28XN
SERVICE BUS SUPPLY CONTRL GRND&FLT AC	124 VU, A6	9XX
115 VAC ESS		
VOLTAGE IND AC ESS BUS	22 VU, H5	15XV
115 VAC EMERG		
VOLTAGE IND AC EMRG BUS	22 VU, H6	16XV
115 VAC FLIGHT-GROUND		
FUEL GAUGE NORM SUPPLY	124 VU, A19	1XR

switch or lever operations. System checks fall into two broad categories: (a) checks of logic and (b) checks of operation. After these areas are confirmed for the normal operation, one can look at abnormal operation and finally, emergency operations. It is in this latter that one can tell whether the simulation algorithm for the system is a good one or not, and by inference, whether the simulator will be a good training device. The less sophisticated simulator manufacturer can often produce a passable imitation of normal, abnormal and emergency operation which will give all the necessary responses to the tests defined in the ATM, but once the tester moves away from these defined tests and looks for second and third order effects, the shallowness of the simulation can usually quickly be found. It is a matter for the purchaser to decide whether such depth of simulation is necessary for training or not, but one must bear in mind the multi-million pound/dollar simulator is probably destined for a 15-year life and who can say what particular drill or operation may be devised for use in seven to eight years time? The proper algorithm and depth of simulation is pretty good insurance that most eventualities will be covered no matter how long the simulator is in use.

The subjective type of check is less easy. The checkout data on which the ATM is based is derived from aircraft design data, test flight data and wind tunnel data. In recent years, this has improved immeasurably, but even so much of the checkout data is extrapolated from or calculated from just a few test results obtained from an aircraft or wind tunnel. It is inevitable therefore for discrepancies to creep in or for some particular aircraft idiosyncrasy to be missed out altogether. Also, it is rare in the case of a dynamic check, for all rates of change to be delineated. Plots of one parameter against another or against a time base are usually available for basic operations, but it is frequently necessary to supplement these with operator experience and this is invaluable even if the experience is not of the same type of aircraft. The experienced pilot or flight engineer can usually state whether the speed at which an instrument moves from one reading to another is likely to be realistic and this type of information is always difficult to define for every check carried out.

This type of subjective knowledge/data becomes most important in the checking of systems and areas which have little objective basis. Flight, motion, control forces, pressurization and sound are all fundamental parts of the simulator which have a high degree of subjectivity. Only experience can really help here even though recent innovations such as control force plots, automatic test guides and spectrum analysis all assist. However, many a user of simulators can tell of devices which match the plots, well within tolerance, but which do not 'feel' like the aircraft being simulated.

The problem is that the industry does not yet know which stimuli (present or otherwise) are of primary importance in making the simulator feel like the aeroplane although everyone in simulation has his own idea.

Subjective checking is essential, but is fraught with danger. Most aircrews charged with an acceptance will have personal experience of most normal operations, but will find it difficult to describe many of the aspects of normal operation, e.g., a crewman can tell easily when the simulator operation is wrong, but often has difficulty in describing what should be happening because, on the aircraft he is trained to look for abnormalities and take the norm for granted. He will frequently spend many hours in flight looking at the 'overall picture' given by an instrument panel display in a certain set of circumstances, he will not easily be able to recall the actual reading of any one individual instrument unless it has a particular significance in the operation. When the testing of abnormal and emergency conditions is involved, the subjective tests become even more of a problem. Each member of the crew believes he knows what conditions will exist for any given emergency situation, but few have personal experience of that emergency. Their store of knowledge comes from books, deductions, hearsay or imagination, all of which sources are frequently incorrect. The worst kind of subjective knowledge comes from experience on another simulator!

Another danger in any checking, be it subjective or objective, is the use of simulation to check simulation. In its simplest form, this means, for example, to insert a simulated malfunction from the instructors' station in order to confirm the systems operation. It is much better to introduce the failure of a power supply by pulling a circuit breaker rather than by operating a malfunction which simulates that power supply failure. In both cases, a degree of simulation is used, but in the former case, the effects of the CB trip are not limited by the malfunction description used in the design of the failure. Some important second and third order effects which on the surface may appear to be totally unrelated are frequently omitted by the system design engineer when defining a malfunction simulation. The design engineer also is easy prey to a misconception of how a particular system operates. Given a limited amount of design data he produces a simulation based on assumptions which coincide with his conception of a system operation which produces the desired results. If the same engineer is also responsible for writing the acceptance test for that system, he will design the test to prove his model complete with its erroneous assumptions and the circle is then closed. It is for this reason, amongst others, that the pre-acceptance testing is best never left solely to the system designer. The better pre-acceptance is carried out by evaluation pilots and engineers who view the simulator and test it in the same light as will the customer team. This presupposes that these evaluation personnel are recruited from experienced pilots or flight engineers and have their experience maintained by periodic training and handling exposure. Only one simulator company has to date followed this plan, but the results have proved the validity of this statement. Ideally, of course, the same evaluation pilot should have been involved in the production of the ATM, serving as quality control of its content and sequence, thus identifying any wrong assumptions in a very early stage of the simulator development and any mismatches between the ATM and the operating procedures defined in the aircraft manuals.

Fig. 9.8 shows the first page of a typical ATM Chapter illustrating the necessity to define the initial conditions. From this, one proceeds to the start of system checks illustrated by Fig. 9.9 which is an extract from a recent ATM.

Although a good ATM is essential as a guide to testing, no manufacturer should expect a customer acceptance team to check only the items delineated in the ATM. The ATM, at best, is a basic guide to ensure primary areas are not left untested, The problem is, however, that to do this task properly usually results in the simulator being tested in conditions which are unlike those in which it will be used when in training. The tests to prove compliance with data have to be conducted in conditions derived from those existing when the data was recorded.

As a result, most testing is carried out in standard ISA conditions with perhaps a repeat at ISA + 10, + 20 or + 30 and again at ISA minus conditions. If the system designer has done his job correctly and has had access to the necessary data, this will normally suffice and the test data for one area will match that for an associated one.

Fig. 9.8. A300 B2K —200 flight simulator acceptance test manual.

2. ON-GROUND CHECKS		COMMENTS	RESULTS
A. Initial Setup			
(1) Set up A/C stationary on ground under the following conditions:			
(a) Altitude	: SL		
(b) Temperature	: 15°C		
(c) Barometric pressure	: 29.92 in Hg		
(d) Parking brake	: set		
(e) Electrical GPU	: disconnected		
(f) Pneumatic GPU	: disconnected		
(g) Engines	: idle		
(h) APU	: running normally		
(i) All other systems	: operational		
(j) All CB's	: in		
(k) ANNUN LT sws	: BRT		
(1) X BLEED VALVE SELECTOR	: AUTO NORMAL		

A customer would certainly be suspicious of a test which proved engine performance at $+30\,^{\circ}$C when the only fuel system tests were at $0\,^{\circ}$C and the pneumatic tests were at $-5\,^{\circ}$C. The preponderance of testing is thus done at ISA plus and ISA minus, i.e., conditions rarely met in the real world. It is important, therefore, for the acceptance team to move out of this straight-jacket now and again to look at more realistic conditions albeit in a subjective way.

It is impossible to test all combinations of system interdependence in all conditions without spending more than the usual six to eight weeks allocated to acceptance so the acceptance team needs to be selective in its choice of testing outside of the ATM and to concentrate on useful testing, if possible based on personal experience.

9.5.6   *Configuration control and defect reporting*

The most exhaustive testing has little value if the defects noticed are not properly reported, rectified and retested. The question in every tester's mind is, however, 'in correcting one defect, how much else is changed – do I need to recheck the whole test, the whole area or even the whole system?' The honest answer is that a compromise must be agreed. Theoretically, the only sure way is to correct the defect and then go back to the beginning and start all over again. This is not practical because the testing would never finish! The compromise must be to check the rectification for immediate effects, load the change into the system and then

Fig. 9.9. A300 B2K — 200 flight simulator acceptance test manual.

		COMMENTS	RESULTS
B.	**Engines Fire Protection System**		

(1) Trip ENG 2 FIRE DETECT LOOP A CB (22VU, C11), ENG 2 FIRE DETECT LOOP B CB (22VU, C12), APU FIRE LOOP A WARN CB (129VU, C16) and APU FIRE LOOP B WARN CB (129VU, C21). Check that no indicator changes for the following positions of the ENG 1 FIRE DETECT LOOPS sw.

    (a) ENG 1 FIRE DETECT LOOP sel @ A   : no effect   :   :   :

    (b) ENG 1 FIRE DETECT LOOP sel @ BOTH   : no effect   :   :   :

    (c) ENG 1 FIRE DETECT LOOP sel @ B   : no effect   :   :   :

(2) With the ENG 1 FIRE DETECT LOOPS sw @ B, press the LOOP TEST A pushbutton. Check:

    (a) All indicators   : no change   :   :   :

(3) Set the ENG 1 FIRE DETECT LOOPS sw @ BOTH, press and hold the LOOP TEST A pushbutton. Check:

    (a) ENG 1 LOOP A lt   : illuminates   :   :   :

    (b) Master Warning System LOOPS lt   : illuminates   :   :   :

    (c) Single Stroke chime   : audible   :   :   :

(4) Set the ENG 1 FIRE DETECT LOOPS sw @ A and hold the LOOP TEST A pushbutton depressed. Check:

    (a) ENG 1 LOOP A lt   : remains illuminated   :   :   :

    (b) Master Warning System LOOPS lt   : extinguishes   :   :   :

    (c) Master Warning System ENG 1 FIRE lt : illuminates   :   :   :

watch for any adverse side effects. Recent advances in computer operating utilities have helped in this respect. It is now possible to have stored on one disk pack three or even four versions of the simulator program, identified as Grandfather, Father and Son files with perhaps a further Work file as well. At the commencement of any test period, all files should be at the same revision level. As testing progresses, errors in the simulation will be found and defect reports raised. These in turn will be rectified by program changes which will be made on the Work or Son file. The test team may then check these changes on this file in isolation. Those changes which give the desired effect may then be loaded to the next level of file which is kept as the current testing file. In this manner, only tested changes are loaded to the file which contains the systems which have already been accepted. Every three or four days, the Test file is loaded to the next file above it, thereby ensuring there is a fall-back load in case an 'accepted' change is found to have unwanted effects a few days later and there is need to backtrack.

9.5.7 *Automated test guides*

These were introduced as a result of the FAA's Advanced Simulation Plan and are as much a test of repeatability and an exercise in time saving as a test of simulation. The theory is that an aircraft of the type to be simulated is instrumented specially for the purpose of obtaining checkout data which can be recorded as a series of time histories. The test pilot flies a number of test manoeuvres, the inputs to which are as carefully recorded as the results. These inputs are then repeated in the simulator which has the means of recording the same parameters as on the aircraft. If the simulation is correct, within the agreed tolerances, the plots from the aircraft may be overlaid on those from the simulator to confirm the simulator performance. In practice, there are several problems, the greatest of which is that it is frequently very difficult to repeat the aircraft pilots' inputs exactly on the simulator with the consequence that different results are obtained. This requires the tests to be repeated several times until an acceptable result materializes. To reduce such problems, there is now a move towards automating the ATG so that the simulator host computer reproduces the original pilot's input, thus obviating the risk of the simulator pilot introducing deviations in the results.

9.5.8 *Summary*

To facilitate the Evaluation and Acceptance Testing and ensure a good simulator:

(1)   prepare a good ATM

(2)   have the ATM approved by the Customer's Acceptance Team prior to start of test

(3)   plan to run the tests in the ATM as the Evaluation Phase and correct any errors which may be found

(4)   maintain strict file configuration control during all testing

(5)   spot check the simulator after its installation using the ATM as a basis and then confirm its final acceptability by using the simulator to fly out typical training flights making use of malfunctions.

10

The flight simulator as a research tool

10.1 Introduction

The design and development of modern aircraft makes extensive use of flight simulation. A vast range of problems is open to investigation on simulators. The essential feature of all such investigations is to introduce the pilot into a closed loop control situation, so that account is taken of his capabilities and limitations. The expectation is that within the bounds of the experimental conditions, his behaviour in the simulator matches his behaviour in the flight situation.

A frequent discussion point is the need in a Research Simulator to duplicate the flying environment. The difficulties of doing so have been dealt with in earlier chapters: wide-angle, high resolution visual display; large amplitude six-degree-of-freedom motion system; complex aircraft modelling and full cockpit fit. But, the argument goes, such complexity is self-defeating; no investment can reproduce fully the flight environment. What is needed is the reproduction of only those stimuli which contribute to the pilot's control task. Furthermore, the stimuli may be exaggerated (in comparison with the real world), to compensate for some other short-coming. In this way, the pilot will be less aware of the lack of realism, and will perform in a manner more closely matched to the technique which he uses in flight. AGARD (1978). One useful outcome of such discussions is that analyses and experiments are recommended, and performed, to isolate the cues which are used by pilots. The value of this work lies not only in helping to set simulator standards for other research work, it also helps to define training simulator standards. Unfortunately, experimental work of this type places considerable demands on the simulator in which it is conducted: to identify the effect of a degradation in cueing, the equipment must have as a starting point a very high standard of fidelity. For example, research on the value of motion cueing, and the use of washout, needs the

Table 10.1. *Research facilities – variable stability aircraft*

Country	Organisation	Location	Aircraft	Work
USA	Calspan	Buffalo NY	Lockheed T-33A Convair C-131H Bell X-22A	Fighter flying qualities Total In-Flight Simulator (TIFS) VTOL research
	Princeton Univ.	New Jersey	Navion	Aircraft flying qualities (6 d.o.f.)
	NASA Ames	Moffett Field	Boeing CH 47B	Helicopter control displays (4 d.o.f.)
	NASA Ames	Moffett Field	D-H QSRA	Powered-lift STOL performance and control
Canada	NAE Ottawa	Ontario	Bell 205	VSTOL Flying Qualities controls/ displays
W Germany	DFVLR	Braunsweig	Hansa-jet VFW 614 BO 105	Flying qualities, displays, fly by wire, fly by light. Helicopter flight control

large motion amplitude systems available only at NASA Ames, or RAE Bedford.

It is important therefore, to choose the right simulator for the research investigation which is planned. Extensive simulation facilities are required throughout the procurement cycle of large aircraft projects. The choice of aircraft configuration will be influenced by early simulator evaluations; trade-off studies will determine the balance of airframe and equipment. At the design phase, aspects of stability and control, and the primary flight control options will be simulated. On other simulators, human factor considerations of the design, such as cockpit layout, controls, displays and crew work-load will be assessed. As hardware becomes available, it is incorporated into engineering simulators for evaluation, and before flight, the simulator will be used for pilot familiarisation. Flight test clearance will be supplemented by parallel activities on ground-based simulators.

An indication of the range of simulators in use for research and development is seen in Tables 10.1, 10.2 and 10.3. They fall into three categories – in-flight simulators, ground-based government or university facilities, and ground-based facilities in industry. The size and capital cost of the ground based facilities varies enormously, from around $100 million to little more than $10,000, reflecting the scope of research work which can be performed on flight simulators.

Table 10.2. *Ground simulator research facilities – government/universities*

Country	Organisation	Location	Type	Size
USA	NASA Ames	Moffett Field, CA	Man/vehicle system research Rotocraft and VSTOL Spacecraft Simulation technology – large amplitude motion	v. large
	NASA Dryden	Edwards AFB	Flight test support, Simulation technology	medium
	NASA Langley	Langley, VA	Differential Manoeuring Simulator (DMS) Displays and flight control concepts Simulation technology	v. large
	Wright Patterson AFB	Dayton, Ohio	Displays and controls Simulation technology System development	v. large
	Air Force HFL	Williams AFB Az	Training effectiveness Simulation technology	large
	Naval Training Systems Centre	Orlando, Fla	Training effectiveness Simulation technology	large
	Naval Air Development Centre	Warminster, Fla		medium
	San Jose State Univ.	California	Human factors	small
	Univ. of California	California	Human factors	small
	Univ. of Utah	Utah	Human factors	small
	Univ. of Illinois	Illinois	Human factors	small
	Perdue Univ.	Indiana	Human factors	small
	Ohio State Univ.	Ohio	Human factors	small
UK	Royal Aircraft Establishment	Bedford	Helicopters, VSTOL, Flight control, Simulator techniques Large amplitude motion	large
	Royal Aircraft Establishment	Farnborough, Hants	Systems, Displays, Air combat	large
	Institute of Aviation Medicine	Farnborough, Hants	Human factors	small
	College of Technology	Cranfield	Flight mechanics	small
	Salford Univ.	Lancashire	Techniques	small
	Southampton	Hampshire	Techniques	small

Table 10.2. *Contd.*

Country	Organisation	Location	Type	Size
France	Flight Test Centre	Istres	Flight test support	medium
	ONERA	Chatillon	Research	small
	CELAR	Rennes	Air combat techniques	Medium
Netherlands	NRL	Amsterdam	Flying Qualities, displays Simulator techniques	medium
	Delft Univ.	Delft	Simulator techniques	small
FRG	DFVLR	Braunsweig	Flying qualities, displays, systems techniques	medium
	IABG	Ottobrunn	Air combat techniques	medium
	DFVLR	Oberfaffenhofen	Flight test support	small
	IAM	Cologne	Human factors	small
Sweden	FFA at KTH (Technical Univ.)	Stockholm	Flying qualities, Systems Human factors	medium
	IAM	Linkoping	Human factors	small
India	ADE	Bangalore	Aircraft design and development Human factors Simulator techniques	large
Canada	NAE	Ottawa	Flight mechanics	small
	Univ. of Toronto	Toronto	Techniques	small
Japan	Univ. of Tokyo	Tokyo	Human factors	small

Table 10.3. *Ground simulator R and D facilities – industry*

Country	Organisation	Location	Type	Size
USA	Boeing	Renton, Seattle	Civil aircraft R and D Flight deck design Simulation technology	large
	Bell	Dallas, Texas	Helicopter R and D	medium
	General Dynamics	Fort Worth, Texas	Advanced fighter design	medium
	Grumman	Bethpage, NY	Aircraft research	small
	Lockheed	Los Angeles	Maritime tactical systems	medium

Table10.3.

Country	Organisation	Location	Type	Size
	Lockheed	Georgia	Flight deck design	medium
	LTV	Dallas, Texas	Aircraft research	small
	Martin	Marietta	Systems development	medium
	McDonnell-Douglas	St Louis, Mo.	Military aircraft R and D Air combat VSTOL	large
	McDonnell-Douglas	Long Beach, Ca.	Civil aircraft research	small
	Northrop	Los Angeles	Fighter R and D Simulator technology	large
	Rockwell	Los Angeles	Military aircraft R and D	medium
	Sikorsky	Stratford, Ct.	Helicopter R and D	medium
UK	British Aerospace	Bristol	Human factors	small
		Brough	Military R and D	small
		Dunsfold	Systems	medium
		Hatfield	Civil R and D, research	small
		Weybridge	Civil displays	small
		Warton	Military R and D Air combat	large
	Westland	Yeovil	Helicopter R and D	medium
France	Dassault–Breguet	St Cloud	Military R and D	small
	Aerospatiale	Toulouse	Civil R and D	medium
		Merignane	Helicopters	small
FRG	Dornier	Friedrichshafen	Military R and D	medium
	Messerchmitt–Bolkow–Blohm	Ottobrunn	Military R and D Helicopters	large
		Bremen	Civil R and D	small
Italy	Aeritalia	Turin	Military R and D	medium
	Aeromacchi	Varese	Military R and D	small
Netherlands	Fokker	Amsterdam	Military and civil R and D	small
Sweden	Saab–Scania	Linköping	Military and civil R and D	large
India	Hindustan Aeronautics Ltd	Bangalore	Military R and D	small
Australia	Government Aircraft Factory	Melbourne	Military R and D	small
Canada	Canadair Ltd	Montreal	Civil R and D	small
	De Havilland (Canada)	Downsview, Ont.	Civil R and D	small

10.2   In-flight simulators – variable stability aircraft

Pioneering work in the development of variable stability aircraft was done at Cornell Aeronautical Laboratory (later to become CALSPAN) in the 1950s. The most successful aircraft has been their T-33A, still flying after almost 30 years. As well as being used for general research in flying qualities, investigations have been carried out on the T-33A, on the behaviour in flight of specific aircraft at the design stage. CALSPAN have also operated a variable stability B-26 for pilot training, a Bell X-22A for VTOL research, a highly modified Convair 131H for total in-flight simulation (Fig. 10.1), and a variable stability Learjet. A fighter replacement for the T-33A is under consideration.

The principle of operation of all these aircraft is to modify the apparent stability and control response of the aircraft by feeding back response parameters into the electrical flight control system, and by shaping the pilot's control inputs. The advantages of experiments performed in flight are self-evident – the pilot is in his natural element, the visual display does not pose a problem, and the correct motion stimuli are available. At the same time, it is not always possible to match the relevant flight condition or flying task. Tests on a variable stability aircraft are less likely to produce the large volume of results which are sometimes needed in research – for example, when a large number of pilots must be used to obtain a statistically significant result. In these cases, ground-based trials are preferable.

Fig. 10.1. Variable stability aircraft: Calspan total in-flight simulator; a converted Convair C-131H.

The specification of desirable flying qualities of aircraft and helicopters has benefitted greatly from tests on variable stability aircraft. (Neal 1970). In addition to those listed in Table 10.1, a modified Mirage III at Istres Flight Test Centre in France made extensive tests in the period 1965–75. Much of the work has been complementary to ground-based simulator trials. With improving standards of ground-based equipment, the need for extensive testing on variable stability aircraft has been reduced.

10.3   Ground-based facilities: government/university

Ground-based research simulators came of age in the 1960s. From a subsidiary role in the fifties, they matured to being an essential part of the aircraft design process by the end of the sixties (Chapter 2). In particular, the success in the US of the space effort, which led to Neil Armstrong's moon walk in 1969, would not have been achieved without the extensive use of flight simulation in the Mercury, Gemini and Apollo programmes. The simulators at NASA Centres allowed astronauts, engineers and programme managers to respond to the challenge. Techniques were developed for docking, lunar landing, lunar surface activities and spacecraft recovery.

The most extensive simulator research facilities are in the US. The scope of testing at NASA Ames and NASA Langley covers the entire spectrum – civil and military, conventional aircraft, unconventional aircraft, vertical take-off and landing, helicopters flight control, displays, systems, man/ machine interface, and simulator technology. Additionally, military interests are also served by the large facilities of the USAF, at Wright–Patterson Air Force Base, and of the USN, at Orlando, Florida. Many developments in simulator hardware have been pioneered there. Western Europe also possesses extensive government funded research facilities which stand comparison with those in the US (Fig. 10.2).

The range of topics which they investigate may be judged from Table 10.2. They are supportive to activities which include military aircraft specification and procurement, civil aircraft specification and certification, and simulator standards for pilot training (AGARD, 1980). Each of these activities calls for funding on a massive scale, often at considerable risk. It is worthwhile therefore to offset some of that risk by research on a large scale. For example, the advanced civil flight-deck research facility at NASA Ames, operational in 1985, cost over $50 million dollars, and allows realistic crew workload assessments to be made of future airline operations from engine start to shutdown. In the UK, a new five degrees of freedom motion system at RAE Bedford (Fig. 10.3) allows cockpit translations of ± 5 m horizontally and vertically. The device will investigate new forms of flight control, and the influence of turbulence on aircraft, crew and passengers.

Fig. 10.2. (a) Cutaway sketch of Man – Vehicle Systems Research Facility at NASA Ames (1984).

(a)

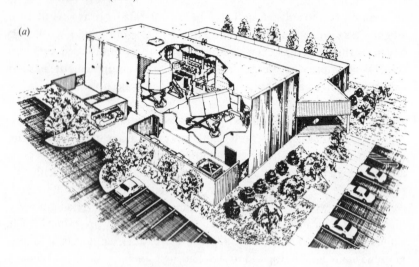

Fig. 10.2. (b) Research simulator at the Netherlands Research Laboratories, Amsterdam, including a computer complex, four-axis motion system, and TV/model board visual system (1975).

(b)

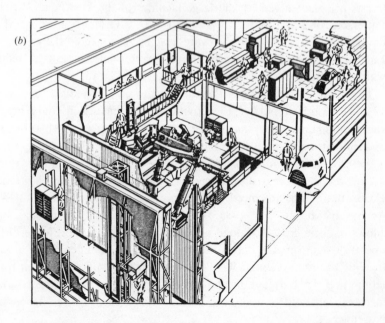

Fig. 10.3. RAE Bedford flight simulator complex.

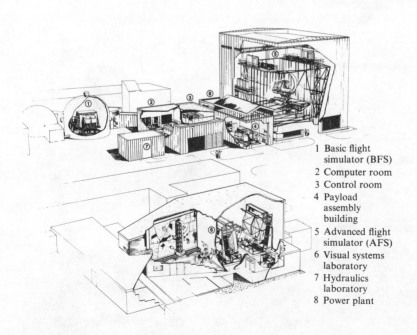

1 Basic flight simulator (BFS)
2 Computer room
3 Control room
4 Payload assembly building
5 Advanced flight simulator (AFS)
6 Visual systems laboratory
7 Hydraulics laboratory
8 Power plant

Many problems in aviation may be solved on a smaller scale; it is in the universities that such efforts are often located, in faculties of aeronautics, engineering or psychology. The same principles apply – a loop closure of control input – aircraft response – display presentation – pilot action. By restricting the scope of the simulator to a particular aspect, the hardware requirements can be minimised. In the limit, useful results are often achieved with nothing more than a microcomputer, a mini-stick, an oscilloscope and a chair (Fig. 10.4).

The combined efforts of government/university ground simulators allows progress on a broad front. The implications of new technology on aircraft design and operation can be assessed. Design specifications can be developed in areas such as flying qualities, flight control, and avionics. The limits on performance imposed by pilot workload, and the improvements possible by applying ergonomic principles can be explored.

10.4   Ground-based facilities: industry

The investment in flight simulation facilities in industry matches that of the research establishments. The distinction between the two lies in their utilisation – the facilities in industry are, by and large, project-oriented. No major aircraft manufacturer would undertake a new aircraft programme without the use of ground-based simulators to help in the

Fig. 10.4. 'Stick and Scope' simulator, Fiat Aviation, Turin (1958).

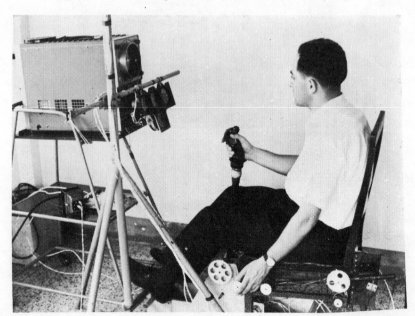

design and development process (Jones 1979). They now rank alongside wind tunnels and system test rigs as an essential part of project definition, engineering and demonstration.

At the project stage, the emphasis of the simulation in aircraft design work is on performance and operational use. Can the potential benefit of a new design feature be fully exploited operationally, when other factors come into play? For example, if high lift devices for the wing are developed, what are the implications on short-field performance? Are flight control systems improvements needed? Will crosswinds be more of a problem?

Similarly, extensive investigations are possible using air combat simulators to quantify the benefits of improved airframe or weapon performance in combat. These benefits can be compared with those that would result from an improved avionic fit. Cockpit layout changes can be explored. Even such basic questions as whether a new fighter aircraft should have a one or two man crew can be investigated by simulation (Zaitzeff 1969).

At the design stage, accurate computer modelling of the aircraft and systems is required. It is only in recent years that the necessary computer capacity and speed has become available to allow real-time modelling of the aircraft and its systems over the full flight envelope. As new data emerges from wind tunnel testing and the flight control specialist, rapid changes to

Fig. 10.5. Tornado flight control system rig, MBB, Ottobrunn (1974).

the model must be possible, calling for the use of a good operating system, with facilities to check that the desired change has been implemented. The simulator at this stage becomes a focal point for much of the design activity, since it provides common ground, on which project managers, engineers, and pilots can resolve many of their problems. The flight simulator has changed the traditional image of the test pilot, who used to be seen as a 'devil may care' extrovert, isolated from design, who expected trouble when the prototype appeared. Now he is part of the project team, he is qualified in engineering, and is disappointed if events in flight do not closely follow prediction.

The simulator used in industry at the design stage will not be fully representative of the project aircraft. As hardware becomes available, it will be incorporated into the simulator. At the same time, rigs to test aircraft equipments will be constructed. It is usual practice now to perform closed loop as well as open loop testing on avionics and flight control rigs, and the distinction between rigs and simulators becomes blurred (Fig. 10.5). A valuable part of rig testing is to make pilots familiar with the operational aspects of the aircraft and its systems, prior to first flight, and in support of development flying. At this stage, the benefits of a flexible computer-operating system, using high level language, must give way to the formal procedures of inspection and test which are applied to any system intended for flight use. It is by adhering to such rules and by using in the simulation many of the actual aircraft components, that a major reduction can be made in the number of unpredicted occurrences in flight.

10.5   Data sources

Central to the successful representation of the behaviour of an aircraft is the quality of the data set used to describe the aircraft (Chapter 6). The exact requirements for data, in terms of scope and accuracy, depend on the particular research programme. If handling or performance matters are secondary to some other aspect, a simple description of the aircraft, using linear, small perturbation modelling may suffice. If the reaction of the aircraft to control inputs is under examination, a better standard is called for.

There are several sources for these data. Calculation or testing can provide weights and inertias; aerodynamic coefficients come from computation, wind tunnel tests, or from the analysis of data obtained in flight. Nowadays it is not unusual to allocate some flight testing to measurements required for updating the model in the flight simulator. The clearance into service of new aircraft depends more and more on the use of ground-based flight simulation. Consequently, emphasis is now placed on providing an accurate aerodynamic description of the aircraft as early as possible, with regular updates from flight test analysis.

10.6   Validation

Although it is impossible to reproduce on the ground all the characteristics of an aircraft as seen by a pilot in the air, the assumption behind the use of the simulator for research purposes is that the pilot controls the simulator in the same way as he would the aircraft. The purpose of the validation exercises is to demonstrate the truth of this assumption.

The validation process can be broken down into components. Data sources can be checked, perhaps by comparing wind tunnel results with the results from flight. The performance of hardware can be compared to specification. The response to control inputs can be calculated independently, and then compared with the equivalent response measured in the simulator. But the final test is the pilot's subjective impression of whether the simulator is like the aircraft. Such conformation is only possible when an existing aircraft is being simulated. When investigating new aircraft, however, it is essential to compare them on the simulator with an existing aircraft so that a pilot familiar with that aircraft can lend credibility to the behaviour of the new one.

The advent of all-digital modelling has helped considerably in achieving good validation. Once established in the computer, the data set is less prone to change due to human error or equipment failure; also if corruption does occur, it is readily detected. Additionally, the computer itself does not limit the accuracy of the model.

10.7   Pilots

In training simulators, the pilot is instructed; in research simulators, he is instructive. He takes a vital role in the process of assessment. Part of that role, as we have seen, is in validation, but the major part is to extrapolate his experience into the flight situation. To do so requires special skills. A sound knowledge both of aeronautical engineering, and of the limitation imposed by simulator hardware, is essential. He must also recognise the context in which an investigation is made, so that the significance of any results can be assessed in terms of their application in aircraft design. Finally, he must be able to transfer his performance in the simulator into flight, making allowance for a different level of workload, and pilot skill. Nor must he underestimate or overestimate the transfer.

10.8   Two examples of research investigations using ground-based flight simulators

The following case studies illustrate the factors which must be considered in preparing an experiment, and the methods which are applied, when using a flight simulator for research.

A Research on flying qualities

A.1. *Background*

Aircraft which are highly regarded by pilots not only perform well, they also possess 'good flying qualities'. What constitutes good flying qualities? Because they cannot be measured purely in engineering terms, they are more difficult to define than structural integrity, or take-off performance.

The stability of the aircraft, and its response to control inputs affect flying qualities, and can be quantified. But the pilot's judgement of flying qualities is made in relation to the task he is performing, so that factors such as the disposition of controls, and the cockpit displays, will have a bearing.

From the designer's point of view, it is desirable to have a clear statement of criteria to be met to ensure good handling, rather than to wait for a subjective assessment. For advanced aircraft designs, commercial considerations call for a set of explicit requirements, to form part of a specification. Such requirements exist for civil and military aircraft designs. Additionally, the flight simulator can be used to establish criteria, to help in the design process.

The underlying principle in using the flight simulator for research on flying qualities is to find a unique relationship between pilot preference on the one hand, and a measurable quantity, such as stick force per g, on the other hand.

A.2. *Measurement*

Pilot preference is a subjective measurement, obtained by the use of a rating scale. The Cooper–Harper scale (Cooper & Harper 1969) is in widespread use, and consists of a ten point scale (Fig. 10.6) which divides into three subgroups – 'satisfactory', 'satisfactory with room for improvement', and 'unsatisfactory'. Within these categories, finer gradations are possible.

Many quantities are needed to describe completely the stability and control of an aircraft. Early efforts at flying qualities research tried to associate pilot rating with changes in aerodynamic derivatives. More success was achieved later using the modal parameters of the stability quartic. Now, compound parameters are commonplace, some of which only appear when the closed-loop stability of aircraft and pilot, rather than the open-loop response of the aircraft, is examined. Nevertheless, the principle of performing trials on a ground-based simulator, and relating pilot opinion to parameters which describe the aircraft behaviour, applies equally to the simple experiments described below, and those based on more complex descriptions.

A.3. *Rolling requirements*

Lateral manoeuvres depend primarily on the ability of the pilot to apply bank angle, through the aileron control. The single degree of freedom rolling mode is characterised by a first order lag between aileron application and the achievement of a steady rate of roll. The lag time constant, τ_R, is inversely proportional to the aerodynamic damping L_p. The initial rolling acceleration of the aircraft, $\dot{p}_M$, is determined by the aileron control power, $L_{\delta a}$. The steady state rate of roll p_{ss} is equal to $L_{\delta a}\tau_R$.

Figs. (10.7 and 10.8) show results from two ground-based simulator trials to establish the relationship between pilot rating, $\dot{p}_M$, and τ_R, for the landing approach task. One experiment studied the requirements for a fighter aircraft, the other experiment provided equivalent results relating to a transport aircraft (Barnes 1967).

Fig. 10.6. Handling qualities rating scale.

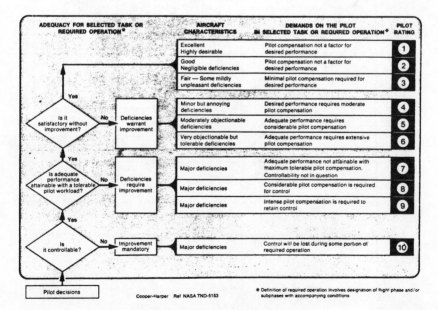

A.4 *Simulator standard*

The closer the simulated task is to that which would be experienced in the aircraft, the less extrapolation is needed by the pilot. For a trial of this type, however, the cockpit does not have to represent a particular aircraft and a linearised model of the aircraft is sufficient. The landing task requires a visual display – these results were obtained using a single window TV/model visual system, with a 1000 : 1 model scale. Pilots were able to fly the approach to touchdown. Simulated turbulence added to the pilot's control task. The experiments were conducted in a fixed base simulator.

A.5 *Experimental method*

Pilots taking part in this type of study must be familiar with the task they are evaluating, the rating scale they apply, and the purpose of the tests. They are not given prior knowledge of the configuration they are assessing. The

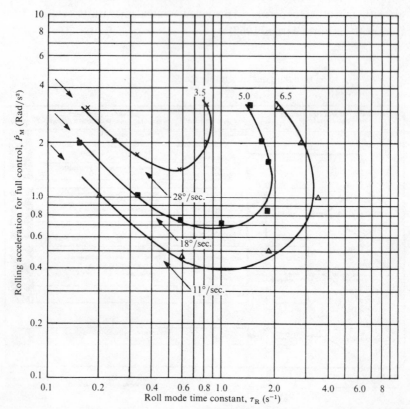

Fig. 10.7. Pilot opinion boundaries: trainer/strike aircraft.

experimental design should ensure that the order in which configurations are assessed does not bias the results, and sufficient repeat assessments with the same pilot, and with other pilots must be included to take out the effects of pilot variability. A typical experiment will use six pilots, and three evaluations of each configuration by each pilot. To produce each of the carpet plots of Figs. 10.10 and 10.11 required sixteen configurations (selected values of $\dot{p}_M$ and τ_R) to be assessed. The plots are constructed by taking the mean pilot rating at each configuration, and identifying lines of 'constant pilot opinion' on the graphs.

A.6 *Results*

Both plots show that for a particular value of damping in roll, handling difficulties occur if the roll control power is either too high (oversensitivity) or too low (unresponsive). Too little damping is very undesirable.

Less obvious is the reason why there should be a difference in the rolling

Fig. 10.8. Pilot opinion boundaries: transport aircraft.

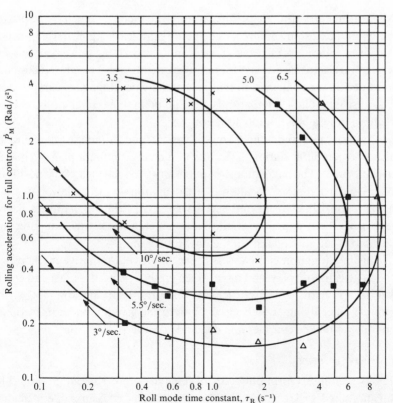

Rolling acceleration for full control, $\dot{P}_M$ (Rad/s^2)

Roll mode time constant, τ_R (s^{-1})

requirements for large and small aircraft: it is clear from these plots that the small aircraft requirements are more stringent. The explanation lies not in the physical dimensions as such. It lies partly in the task, partly in the longitudinal response of the aircraft, and partly in the different response to turbulence. Pilots expect higher manoeuvrability from fighter aircraft; also the higher inertia of large aircraft reduces the control corrections needed to compensate for lateral gusts.

A.7 *Comment*

The rolling behaviour of known aircraft can be easily calculated and plotted on the appropriate diagram, to provide confidence in the experimental results. Once confidence is established, the curves allow rapid inspection of the influence of configuration or control system changes on the behaviour of a new aircraft design.

B Research on advanced avionics

B.1 *Background*

Modern military aircraft performance is enhanced as much by improvements to the equipment fitted, as by aerodynamic or engine developments. Airborne computers increasingly offer more processing power; sensors for input, and electronic displays for output also improve. The aircraft designer must decide the contribution that new avionic developments can provide. The research flight simulator is helpful in making such decisions, by evaluation in a closed-loop situation.

A typical case is the comparison of three methods of entering data into an airborne computer during flight, due to a change in flight plan. (White & Beckett 1983). The navigation computer must be informed of the positions of revised waypoints, to allow the calculation to be make of steering information, time to go and fuel usage. The string of numbers which represent the waypoints is normally entered into the computer by keyboard, suitably placed on the cockpit coaming.

This investigation compares the merits of two alternative methods – a touch-sensitive display, and direct voice input. The need for fast and error-free insertion of data is particularly important for the low level strike aircraft, since flying at high speed, close to the ground, under operational conditions makes considerable demands on the pilot's skill.

B.2 *Measurement*

Pilot opinion can form only part of this type of assessment. It should be substantiated as far as possible by performance measures. Performance measurements in this context are difficult to make. One difficulty is the well-known effect that the pilot often works harder with the poorer system and the performance measure is unaffected. At the same time the difference may not be perceived in the pilot comments, because other factors dominate the work-load assessment. Choice of parameters is therefore critical; also the task should be chosen so that any changes to the measurement parameter is attributable to the variables under investigation.

In this example, performance was measured in three ways – by measuring height variation during data entry, by measuring time taken to enter a data string, and by the time that the pilot spent looking into the cockpit during data entry.

B.3 *Simulator standard*

A fully equipped strike aircraft cockpit was simulated, fixed base, with a three window CGI display. The standard was sufficient to allow a low level navigational task to be carried out, involving two legs, and the update of a waypoint store on each of the legs.

Details of the keyboard (thumb operated) and its read-out are seen on Fig. 10.9. The touch-sensitive display (Fig. 10.10) is similar in appearance, is finger operated, but lacks the tactile feel of the keyboard. However, editing – for example correcting a mistaken entry – is more easily achieved. The direct voice input system has a limited vocabulary, and had to be trained to recognise a particular voice. The entered data string could be read on the head-up display (Fig. 10.11).

B.4 *Experimental method*

Eight operational pilots from RAF Squadrons were used for the trial. Each pilot flew four missions with each of the three systems. The order in which

Fig. 10.9. Keyboard unit.

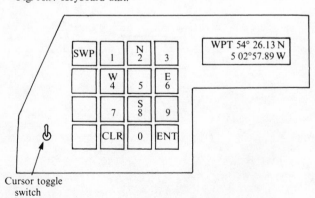

Cursor toggle
switch

the systems were flown was randomised to eliminate learning effect. The latitudes and longitudes entered were chosen to randomise the content and order of the digits. Subjects wore flying suit, helmet, mask and gloves. Each pilot received sufficient briefing and training time to become familiar with each system. Experimental runs were not made until the speech recogniser had been re-trained to suit particular pilots.

Fig. 10.10. Touch-sensing data entry display showing the cursor on the upper readout line.

5	4	°	2	3	.	2	0	N
0	1	°	1	9	.	1	3	W

	N			1	2	3
W		E		4	5	6
	S			7	8	9
Enter	Clear	Set WP	2	0		

Fig. 10.11. HUD showing DVI readout.

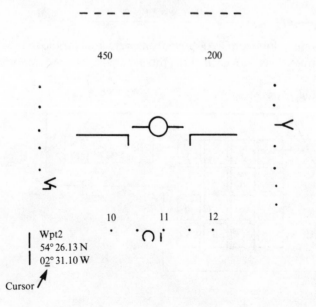

450   ,200

Wpt2
54° 26.13 N
02° 31.10 W

Cursor

Pilots completed a questionnaire after the trial, to allow the subjective impressions to be compared with their measured performance.

B.5 *Results*

Height keeping performance is seen on Fig. 10.12, as a cumulative frequency plot. The deterioration in flying performance during data entry is significantly smaller for DVI than for the two manual systems, illustrating

Fig. 10.12. Comparison of changes in height-keeping performance due to data entry, at 100 ft/550 kn.

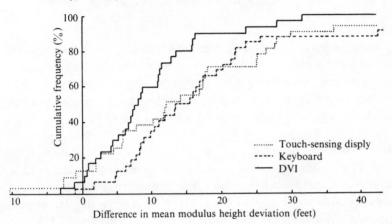

Fig. 10.13. Comparison of head-down times.

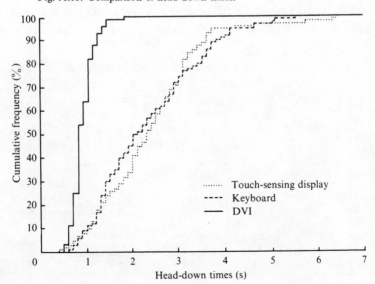

the reduced need for the pilot to look into the cockpit, rather than at the head-up display and outside world. Confirmation of this reasoning is seen on Fig. 10.13, which shows the cumulative frequency of the time the pilots spent looking into the cockpit. These results came from analysis of Eye Mark Camera recordings.

On the other hand, the time taken to enter data via the DVI system was longer (Fig. 10.14). This result is due in part to the recognition ability of the system, rather than due to the pilot's inability to speak more rapidly. Also, because the speech entry was not 100% reliable, pilots waited to check each entry on the head-up display, rather than insert a string.

B.6 *Comment*

This investigation illustrates the pattern of a typical research investigation concerning airborne systems. Although a clear winner did not emerge, the advantages and disadvantages of each system were highlighted, and recommendations emerged for improvements to data entry methods which would have been hard to find without a trial of this kind.

10.9   Concluding remarks

Flight simulation is a vital part of aeronautical research. The research is conducted on a broad front, both at research establishments and in industry. The more sophisticated activities require massive capital investment. Useful results can also be achieved on simple low-cost simulators.

The justification for the investment lies in the contribution which research simulators can make to the design, development and operation of all types of aircraft. Simulation at the design stage reduces the risks

Fig. 10.14. Comparison of total data entry times at 100 ft/550 kn.

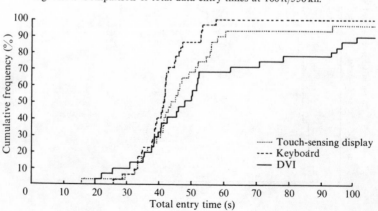

associated with new configurations. The development phase is accelerated by simulator testing and the clearance of a new aircraft into operational use is eased. Results from research simulators also help to improve the standards of simulation used in aircrew training. They prove new concepts and equipment, prior to airborne use and allow evaluation of operating procedures.

The use of flight simulation in research has extended considerably in the last twenty years, as equipment improvements have become available. Significantly, this growth continues, and there are avenues yet unexplored.

11

The flight simulator as a training device

11.1 Introduction

The operation of modern civilian and military aircraft demands very high levels of skill and knowledge. The flight simulator has shown itself to be capable to making a contribution to the task of aircrew training at three levels namely, skill development, skill maintenance and skill assessment.

An example of the significant role that flight simulation plays in training is taken from an overview of the operations of American Airlines (Houston 1984):

> American Airline is a major US carrier. The equipment types flown are Boeing 747, 767 and 727 and McDonnell Douglas DC-10 and MD-80. The total number of airplanes is currently 246 with a total of 259 planned for the end of 1984. Over 1000 take-offs are made each day on the AA system and over 800 000 hours are planned to be flown in 1984. To man this flying operation a total of 4212 pilots and flight engineers are on the payroll.

> Training of all cockpit crews is conducted in a large centralised 'Flight Academy' located near the Dallas Forth-Worth airport in Texas. The ground training facilities include a total of nine flight simulators. The volume of American Airlines training at the Flight Academy in 1983 totalled 7366 training cycles. In 1984 the projected training is expected to be 8894 cycles.

> At American Airlines we are fully committed to total simulation. All transition (conversion) training is accomplished in the simulator with no actual airplane time until the individual's first line flight for operating experience. After the formal rating in the simulator one four hour period is devoted to line-oriented flight

training (LOFT) and one period devoted to intensive take-off and landing practice. Experience shows that this total simulator training fully prepares the individual for his first line trip under supervision, with no exceptions. For the B727 and B767 simulators with daylight–dusk–night, four window displays are used. Captain upgrade and all recurrent and requalification training is also conducted totally in the simulator. Between 1981 and 1984 a total of 9862 aircrew have undergone total simulator training to meet transition, requalification, recurrent or upgrade requirements.

The above statement gives an idea of one airline's commitment to flight simulation in training. American Airlines are not alone in using flight simulators. An appreciation of the extent to which simulators are used for training can be obtained from regular reviews compiled by aviation journals such as Flight International. Their most recent simulator census indicated that more than 100 airlines throughout the world operate training simulations.

In military aviation a similar acceptance of the value of flight simulation has taken place. One estimate has been that the air forces of more than 40 countries use simulators with the US having the highest commitment to their use in training. The military field is particularly noteworthy for the range of simulators that is in use. This reflects the broader and deeper nature of the military aircrew skill development and maintenance task. Simulators are used as part of *ab initio* pilot training (an aspect of training not usually undertaken by commercial airlines), advanced training and conversion to type and specific role, for example, air-to-air refuelling, air combat and airborne surveillance. Once again the aviation journals provide regular reviews of developments in military use of flight simulators.

In the civilian world *ab initio* pilot training is usually undertaken in a general aviation context. Because operating costs are lower and there is a very genuine requirement that training should take place in the air in order to expose the student to the real environment of flight, the widespread use of flight simulation has not been adopted. Nevertheless, simulators have been used in basic training particularly in aviation training schools operated by universities in the US. One example is the Institute of Aviation at the University of Illinois. In such an environment training and research into the application of new methods of instruction can be undertaken. Between 1969 and 1978 a total of 5426 students received flying training, with an average of 542 students per year. In 1969 some 12% of instruction was given in flight simulators but by 1978 this figure had risen to 20%. For a fuller account of this work see University of Illinois (1979) and Roscoe (1980).

The fourth area in which flight simulation has a training function is astronautics. From the initiation of a manned space programme it was recognised that no other means of astronaut training existed other than simulation. The US National Aeronautics and Space Administration (NASA), in collaboration with the simulator manufacturers, therefore embarked upon an ambitious and extensive programme to develop training devices. The Mercury, Gemini and Apollo missions were supported by a wide variety of training simulators. By Apollo Mission 15 59 crewmen had been trained and the total simulator time stood at almost 100 000 hours. Continuing on from these programmes simulators play a key role in the Shuttle and Skylab ventures. For a description of the complexity of simulation for the Shuttle see Bird (1978).

A feature of the NASA simulation program has been the commitment to a 'total simulation' paradigm in which the objective has been to represent as much of the mission as possible. The complexity and accuracy of the facilities required to meet this objective repaid the investment when it was possible to use the simulators to devise recovery procedures for the ill fated Apollo 13 mission. The explosion of a liquid oxygen tank when the craft was 200 000 miles out in space presented an unforseen challenge to the teams back at Mission Control. Using a link up of simulators new procedures were worked out and a successful return to earth was achieved (Kelly & Parke 1970).

11.2   The advantages of simulation

The above introduction provides an overview of the extent to which flight simulators have become accepted as training devices. The applications described have a number of basic differences but there are also a number of common features which indicate why it is that flight simulators have found acceptance in the context of aircrew training. Some advantages were referred to in the first chapter of the book. They now warrant expansion.

11.3   Cost

Aircraft capital, operating and maintenance costs are high. To use an aeroplane for training can incur added costs. For example, training in emergency procedures such as aborting a take-off may involve additional maintenance and replacement component costs. In military aviation the practice firing of guided missiles and other armaments is so costly that very tight restrictions have to be imposed.

The use of flight simulators for training can result in considerable cost reductions. Orlansky (1985), having examined a large number of training programmes using flight simulators, concluded that, simulator acquisition costs are 30–65% of those of the aircraft simulated, operating costs are about 8% of the aircraft and life-cycle costs 65%. If used effectively in training amortization can be achieved over two years.

11.4 Safety

In the simulator potentially hazardous situations can be experienced and dealt with in the knowledge that there is no risk to life or equipment. It is a bizarre statistic that, before the adoption of flight simulators as suitable and effective devices for emergency training, in some situations, more accidents occurred while practising an emergency than resulted from actual occurrence of the emergency (Gillman 1969). The most documented example of this is the loss of the aircraft while practising asymmetric handling after a simulated engine failure (Johnson 1968).

In military aviation, where many high performance aircraft are single seat designs, problems of safety occur when converting new pilots on to the type. One solution is the costly development of a two-seat training version of the aircraft, which may have different handling characteristics from the original. The alternative is to use a simulator which can provide intensive conversion training which is both valid and safe.

11.5 Opportunity

For training to take place in the air, aircraft and airspace has to be available and weather conditions must be suitable. In military training no potential enemy is likely to agree to allow an adversary the opportunity to practise strike missions. These barriers to the adequacy of training can be overcome by the use of flight simulators.

11.6 Ecology

From the mid 1970s world fuel prices have risen and at the same time conservation and environmental protection movements have emerged. The result has been an increase in efforts to reduce operation costs and demonstrate that aviation is concerned about pollution and noise. Training operations can impose high levels of disturbance as aircraft repeat manoeuvres often with an extended operating day. The use of flight simulators can reduce environmental disturbance and save precious fuel resources.

It must be recognised that these pro-simulator arguments can also be applied to any alternative form of training which does not rely upon the use of aircraft. This is an important point for in some training situations devices other than flight simulators can be the more effective means of training. The use of a flight simulator is appropriate when:

(1)   a training task requires a high degree of representation of the actual aircraft, its operating systems and tasks

(2)   the tasks to be trained approach those encountered in the total operation of the aircraft

(3)   the training task involves the assessment of operating proficiency as well as its development.

   Given training situations which meet the above criteria the benefits of simulation expressed earlier can be enhanced by more control over the training task and the training environment, the opportunity for deeper and more objective assessment of performance, and greater flexibility in terms of the ability to vary the content, order, repetition and timing of training elements. For a fuller discussion of these points see AGARD (1980).

11.7   The effectiveness of simulation

   With the potential advantages of simulation over flight as an environment in which to undertake training the question which has to be asked is Do Simulators Train Effectively? Answers to this question are of vital importance to four parties namely:

> The Manufacturer: who must produce a device which meets training requirements

> The Customer: who wishes to meet his training task more economically and effectively

> The Legislator: who is responsible for specifying what training is permitted in simulators substituting for aircraft

> The Training Technologist: who must assist the manufacturer, customer and legislator by conducting research to provide information to aid their decisions

   The function of the manufacturer will not be taken further in this chapter as the remainder of the book addresses itself to the problems of designing and constructing flight simulators. However, the role of the customer, legislator and researcher will be considered more fully.

11.8   The customer's role

   Despite being less costly to operate than an aircraft a flight simulator remains a major item of expenditure. Moreover, as with the aircraft it simulates, it only justifies its purchase when it is in use for training. Rapid amortization of the cost of the simulator can only be achieved if it is in use every day of the week and for 80–90% of each day. Utilisation figures of this order are commonplace with commercial airlines operating simulators. First and foremost the customer must define his training requirements. This task will need to cover both the content of training, i.e. what has to be trained, and the extent of the training, i.e. who has to be trained. The analysis of the tasks to be trained will provide an indication of whether a flight simulator is required, while the determination of the extent of the training task will indicate whether the ongoing requirement in student numbers warrants the acquisition of a simulator.

If the answers to both questions are yes, the customer can proceed with the purchase of the simulator with some confidence. Similarly if both answers are no, he can look for alternative more appropriate methods of training.

The dilemma which may face an operator is that where there is a requirement in terms of task but where the extent of the task is insufficient to guarantee adequate utilisation. In such a situation it may be more cost effective to arrange for training to be carried out by buying time on another operator's facilities. This need is recognised and in commercial and general aviation fields a number of flight simulator training facilities have been set up for the purpose of selling training time to operators who have identified that their training requirements do not justify ownership of a simulator.

It is crucial to undertake a detailed examination of the training task with the aid of experienced operational personnel, training specialists and the aircraft manufacturers. From this study should come a statement of the skill and knowledge that trainees will have to gain in order to operate the aircraft. There then needs to be identified the skill and knowledge levels of the students who are to be trained. For example, is it the intention to take experienced operating crews and convert them on to the new type or will less experienced personnel be trained to operate the aircraft. Relating these two items will identify the training tasks.

The next stage will be to determine the pattern and location of training. A simulator will not be the best training environment for every aspect of the course. Some parts of the training may lend themselves to classroom instruction or self learning packages. For others there may be no susbstitute for in-flight training. The resulting training course is likely to be made up of a number of elements of which the flight simulation phase will be a part. Using a term coined by Eddowes & Waag (1980) the objective is to produce 'least-cost, sequenced, multi-media training'.

The definition and apportionment of training objectives and tasks can be carried out using techniques such as Instructional System Design (ISD) or the Systems Approach to Training (SAT). For a detailed description of these techniques see Miller, Swink & McKenzie (1978) and Cream, Eggemeier & Klein (1978).

Having identified those portions of the training programme for which flight simulator training appears to be appropriate the next step is to produce a specification for the device. At this point the customer must take note of any legislation which sets out the levels of simulation required in order that particular training tasks can be conducted using a flight simulator as a substitute for in-flight training. These important functions are considered more fully in the next section.

In addition to identifying the operational requirements for a flight simulator the customer must decide upon the instructional philosophy to be adopted. Two issues are important and have a bearing on the design of the simulator. They are the choice of instructional personnel for the simulator and the extent to which the simulator facility will have a role in the management of training.

A decision has to made on whether instruction is provided by personnel who will work only in the simulator or whether by personnel who also instruct in the air. Both methods of instruction have their advantages and disadvantages.

For full-time simulator instructors:

(1)   in *ab initio* training it may be wasteful of a flying instructor's time for him to be used to teach procedural drills and cockpit familiarisation in a simulator

(2)   in advanced and operational training the use of full-time simulator instructors can remove the requirement to increase the establishment of qualified line operators to compensate for the time they spend instructing in the simulator

For using flying instructors:

(1)   in *ab initio* training it provides greater continuity as the student receives instruction from the same person in the air and in the simulator

(2)   in advanced training instruction is given by operators who are current on the aircraft and its operating procedures. Thus the credibility of training is not endangered

The design of instructional features and methods of operating the simulator will depend upon the type of instructor who uses the device. A full-time operator may be expected to become more technically acquainted with the simulator whereas the part-time operator will need a more 'user friendly' system, achieved through the design of the simulator or by the provision of full-time workstation operators.

Whatever decision is made the simulator instructors will require training to ensure that they can perform their functions effectively. For a fuller examination of flight simulator instructor training see Royal Aeronautical Society (1977).

Consideration must also be given to the extent to which the simulator will have a part in the management of training. The flight simulator is an example of a device which provides computer based training (CBT). Earlier it was explained how the simulator computer acts as the host to a series of mathematical models and manipulates them to create the impression of flight. Because this process is carried on by using numerical representations of flight and because the computer is well suited to storing and processing numerical data, it follows that the facility exists for recording students' performance in the simulator, analysing it and maintaining a record of standards achieved. If the computer is also programmed with specified criteria it can also compare performance against these and take decisions about its adequacy. A further order of involvement is that if the computer also holds details of the training syllabus and the performance standards expected of a student in order to progress to the next stage of training, it can control the passage of individuals through training. Thus the system now has an element which undertakes computer managed training (CMT). This feature of simulator training is probably one of the most hotly debated of all. There is some resentment of the idea of performance analysis by computer, particularly if the renewal of a licence or rating is linked with the assessment. Moreover, there is some apprehension about the ease with which it is possible to identify valid and reliable measures which can provide accurate assessments of performance and proficiency. Nevertheless, in military aircrew training some success has been achieved. For example in the US Army the basic flying training of helicopter pilots at the Army Aviation Center, Fort Rucker makes use of the Synthetic Flight Training System (SFTS) for the UH1-1H helicopter. On the SFTS students' performance is monitored by the computer which can assign new training tasks and keep a record of performance. Automated performance monitoring is also in use with air combat simulators (Barnes 1984) and air-to-ground attack trainers (Hughes *et al.* 1983). For an assessment of the prospects for computer managed flight simulator training see Dickman (1983, 1984).

The third category of decision that the customer must make is with regard to the policy he will adopt for the acquisition and maintenance of the simulator.

Simulators and aircraft are manufactured by different constructors. So, it has been the established practise for the customer to order the aircraft and simulator separately. This method of procurement permits the choice of aircraft and simulator to be taken independently. However, in order to produce an accurate simulation the manufacturer must rely upon the aircraft manufacturer for extensive design and performance data. Under such procurement arrangements the customer can become a heavily committed intermediary between the two parties.

An alternative policy is to treat the procurement of the aircraft and its attendant training devices as one package and to make the aircraft manufacturer responsible for the provision of the total system. For a description of how one aircraft manufacturer has approached the task of providing training devices to support his aircraft see Martz (1984).

On the subject of maintenance, it has already been pointed out that in order to obtain value for the capital invested in the simulator high levels of utilisation are likely to be necessary. If malfunctions occur maintenance personnel must be available. The customer has to decide whether to employ his own staff or to use contractor maintenance provided by the simulator manufacturer.

To sum up, the customer must identify his training task and the role that a flight simulator will play in meeting the task. He must then produce a detailed specification for the simulator which must take into account the legislative requirements the device must meet. He must also specify how the device will be operated and maintained.

The above description of the customer's task has avoided drawing comparisons between the civilian and military customer. There are differences, for example, the military specification is likely to be more extensive and include unique 'one off' features. As a consequence military procurement costs tend to be significantly higher than their civilian counterparts. For a detailed consideration of the differences between civilian and military simulator procurement practices see Hussar (1983).

11.9   The legislator's role

In civil aviation, in order to have accepted commonality of requirements between countries, recommended standards for the training and certification of flight crews operating commercial aircraft are formulated and published by the International Civil Aviation Organisation (ICAO). These are adopted by national aviation authorities and integrated into their aviation law and regulations.

National authorities are responsible for ensuring that flight crews are adequately trained and that their competence is regularly checked and certificated. Typical of such national authorities are the Civil Aviation Authority (CAA) in the UK and the Federal Aviation Administration

(FAA) in the US. These bodies publish their respective requirements in the form of CAA Civil Air Publications (CAP) which stem from the Air Navigation Order (ANO) and FAA Federal Aviation Regulations (FAR).

Once it had been demonstrated that flight simulators were effective as alternative training environments to the aircraft it was necessary for national authorities to extend their regulations to cover the use of simulators. In the UK the first fixed wing flight simulator to gain approval was a Stratocruiser built for British Overseas Airways Corporation in 1959. The first helicopter simulator to gain approval was an S 61N built for British Airways Helicopters in 1979.

Initially simulators were approved for a limited number of elements with the remainder being conducted in an aircraft in flight. As simulator fidelity improved the number of elements capable of being undertaken in a simulator has increased. Type conversion and revalidation taking place totally in the simulator was achieved in the US in 1981. However in the UK, even when using the most sophisticated flight simulator for conversion to type, the requirement remains for about one hour of aircraft flight time.

The CAA requirements for flight simulators are set out in 'The Approval of Flight Simulators for the Training and Testing of Flight Crews in Civil Aviation' (CAP 453). The FAA requirements are contained in 'Aircraft Simulator and Visual System Evaluation And Approval' (FAR 121-14H).

There is extensive agreement between legislative authorities as to the broad basic requirements that a training simulator must possess. These are as follows:

(1)   the flight simulator must represent closely the full scale flight deck of the aircraft. All controls, switches, instruments and indicators must appear to be identical to those in the aircraft and operate in the correct sense. All systems, e.g., aircraft systems, communication systems, radio aids, area navigation equipment (including Omega and INS), flight management systems, autopilots, GPWS and performance computer systems must appear to function in the simulator as they do in the aircraft. Flight instruments must represent those of the aircraft both statically and dynamically

(2)   a visual flight attachment, that faithfully represents the instantaneous field of view presented to the pilots on the flight deck of the aircraft

(3)   a computer and associated software program, based mainly on aircraft data established during aircraft type certification and supported by data gathered during post certificate flight

(4)   a motion system capable of reproducing realistic accelerations associated with all phases of flight

(5)   means by which the instructional staff can control the conduct of training and tests, inject relevant fault situations and record the results of tests for debriefing and reference

(6)   a technical support organisation and spares support capable of maintaining the simulator to the standard required by the approval

(7)   training arrangements for instructional and technical staff who are employed on simulator instruction testing and maintenance

(8)   full Flight Documentation appropriate to the Company Operations Manual

(9)   a defect reporting and recording system equated to the aircraft documentation and used in conjuction with a Minimum Equipment List (MEL)

There is also an understanding that a simulator, meeting these requirements, should be capable of representing the following normal, abnormal and emergency aircraft handling characteristics:

(1)   internal pre-start, start up, pre-taxy and pre-take-off operation and procedures. There should be visual representation of taxiways, runways and airfield environments. The motion system should provide physical sensations of motion related to speed and taxy surface

(2)   take-off in normal, abnormal and emergency conditions, with visual cues realistically simulated in all conditions. Accurate simulation of aircraft ground speed and ground run to both point of rotation and emergency accelerate stop during take-off.

(3)   climb and cruise in normal, abnormal and emergency configurations, to operating altitudes

(4)   en route line operations between departure airfield and typical destinations and alternate airfields

(5)   approach and landing at designated point of final landing or intermediate landing, (including missed approach procedures) in normal, abnormal and emergency configurations

(6)   approach, flare and landing in normal and all abnormal configurations and conditions with matching ground speed and distance for each condition

During simulated flying the flight simulator must be capable of reproducing representative turbulence, wind effect, icing, reduced visibilities and other meteorological phenomena associated with typical air masses. These must include the visibilities experienced during line operations and during the take-off and landing phases of flight.

With experience of assessing and approving flight simulators the regulatory authorities now acknowledge that different levels of training and certification can be achieved acceptably with simulators of differing degrees of complexity. As a consequence regulations defining the standards of simulation required to meet particular training and assessment objectives have been formulated together with indications of the complementary amount of flight time that is deemed necessary. Standards of this nature have to be arrived at through research and experience and be implemented only after extensive consultation with simulator manufacturers and operators.

For descriptions of the process involved in the formulation of legislation see Ferrarese (1978), Wooden (1978) and Huettner (1984). The CAA propose four levels of approval; Levels 1, 2, 3, and 4. Levels 1 and 2 are appropriate to basic instrument flying training and the instrument rating revalidation test, and Levels 3 and 4 are appropriate to the more advanced flight simulators used for mandatory operators checks and 'total flight simulation' (TFS) conversions. Level 3 flight simulators are intended for use in relation to mandatory operators checks, and currently approved type rating items for pilots and flight engineers. Level 4 flight simulators are more sophisticated, and are intended for aircraft type conversions entirely by the use of flight simulation.

The FAA also have four levels of simulator approval:

> *Phase* I is the landing approval program, for fully qualified air carrier pilots
>
> *Phase* II permits transition and upgrade training and checking
>
> *Phase* IIA permits any part 121 operator to conduct phase II training in a simulator approved for the landing manoeuvre under phase I, if the operator meets additional requirements which are specified by the FAA.
>
> *Phase* III is designed to allow all but static aeroplane training, the line check and operation line experience to be conducted in an advanced aeroplane simulator.

In many respects the technical requirements specified for the CAA Level 3 and 4 flight simulators are similar to those of the FAA Phase II and III. This is quite intentional since it eases the task of the operator and manufacturer in demonstrating performance standards both during initial validation and subsequent operation.

11.9.1 *Evaluation procedures*

In order to ensure that flight simulators intended for aircrew training and certification conform to the regulations, evaluation and approval procedures have to be designed and applied.

It can be considered a tribute to modern simulators that the policy employed by the major legislative authorities is to evaluate simulators in the same way as the real aircraft. The practice normally followed is to conduct an initial evaluation prior to the introduction of the simulator into the training programme and subsequent regular revalidations for the renewal of certification.

The CAA initial simulator evaluation is carried out in two parts, the first being a technical inspection by a software specialist, this is followed by a flight evaluation conducted by a team which includes a CAA Airworthiness Test Pilot with a Flight Test Observer and a Training Inspector.

The technical inspection is carried out to examine, identify and record the standards of modification of software and software documentation. The standard of software so recorded is held by CAA and any subsequent changes are required to be notified and copies submitted.

The Test Pilot then carries out the first part of the flight evaluation programme which is for the purpose of assessing and analysing simulator/aircraft relationships in the following respects: flight deck layout, handling and performance, functioning of flight deck equipments, fidelity and performance of automatic flight control systems, compatability of visual attachment and motion system with the simulator's dynamic performance and the simulation of noises, buffets and crosswinds. This part of the inspection usually follows the same flight profile as would be flown in a real aircraft for certification purposes; real-time data resulting from actual Airworthiness Division flight tests being used for quantitative and qualitative assessment.

The second part of the flight evaluation is carried out by a Training Inspector who is qualified on the specific aircraft type. With full knowledge of the Test Pilot's findings he examines and reports on the simulator's suitability to fulfil the mandatory training and testing requirements for pilots and flight engineers, including the fidelity and scope of all the approach, navigation and communication equipment, the instructor facilities, its suitability for simulator instructor/examiners training and the acceptability of the accuracy of visual displays. The first part of the inspection is flown to evaluate the radio aids en route and at the major airfields used by the operator and it is followed by a typical check/training detail as if it was being carried out in the aircraft. The simulator is evaluated against the Inspector's knowledge and experience of results achieved in the real environment.

In addition to the flight evaluation the Training Inspector inspects and reports on the adequacy of all the supporting facilities including all documentation and safety arrangements.

A Simulator Approval Document appropriate to the finding of the total evaluation specifying the extent or limitations of the approval is then issued. After initial approval has been granted, a type qualified Training Inspector is assigned to the simulator, who keeps the simulator under constant review and submits regular reports on its performance. Each year the approval has to be renewed, the team revisits the simulator and repeats the evaluation and assessment programme. Whilst the inspection and approval procedures by the FAA vary in detail from those of the CAA the content is broadly similar.

With the space available it is only possible to give a very broad outline of the regulators' role in the use of flight simulators. Most countries follow the procedures of either the FAA or CAA which in themselves have become remarkably similar. Fortunately the circle of simulator manufacturers, simulator operators and regulators is relatively small and works closely together.

International seminars on the subject, organised in particular by the Flight Simulation Group of the Royal Aeronautical Society, are excellent examples of this and participation in such events is strongly recommended.

For further more detailed information on the legislators' role and their requirements the following contacts are recommended:

Chief Inspector Flight Operations
Flight Operations Inspectorate
Civil Aviation Authority
Aviation House
129 Kingsway
London WC2B 6NN
National Simulator Program Manager (NSPM)
Department of Transportation
Federal Aviation Administration
800 Independence Avenue. SW
Washington, DC 20591

11.10 The role of the training technologist

The flight simulator designer and the training technologist can at times appear to be on opposing courses. The simulator designer's ambition is to create the most realistic possible representation of the real world. The training technologist seeks to define how to meet the training requirement in the most cost effective manner and to verify that training does take place. For a description of how these aims can be reconciled, together with specific examples from the work of the McDonnell Douglas Corporation see Jones (1978).

The process of identifying training requirements has already been examined but it is important to stress that a knowledge of what actually is the training task is essential if training devices are to match the task. Beneficial training can be achieved using low cost devices ranging from self-instructional programmed books, for example, Culver's 'The Pocket Simulator for IFR Procedures' (FIP Publications), cardboard cockpit facsimiles (Rolfe & Riglesford 1984) and computer based part-task trainers (Knight 1982).

The second aspect of the technologist's task is to measure the effectiveness of training. There are deceptive indications of the training value of a simulator. Reports that a simulator is liked by its students, is considered to be very realistic and is well utilised are by themselves not reliable measures of the device's training effectiveness (Rolfe & Caro 1982).

More acceptable and valid criteria of training effectiveness are not difficult to formulate, namely:

> Can it be shown that, as a result of the use of simulation, students' performance on the actual task is improved, or that specified levels of on-the-task performance can either be achieved more quickly, or with less resort to more costly training resources?

The fundamental concept inherent in flight simulation is that of Transfer of Training (TOT). TOT is the ability for a skilled behaviour which has been learned in one situation to be carried over to another. When the learned behaviour facilitates performance in the other environment positive transfer is said to have taken place. However, in some circumstances negative transfer can occur when behaviour learnt in one situation degrades performance in another. The fundamental principles of TOT were formulated by Osgood (1949). They are that the degree of transfer achieved relates to the extent to which similarity exists between the stimulus and response demands of the training and task situations. For a review of the implications of these ideas to training using flight simulators see Blaiwes, Puig & Regan (1973).

The amount of training transfer can be expressed by measuring how much time spent in the air, training students to a specific performance standard, can be saved by the use of a flight simulator. Note that the task of defining a specific performance standard may not be an easy one and a decision may have be taken as to which of a number of possible criteria of training achievement will be employed. Hammerton (1977) distinguishes between Savings measures, which are based on the reductions in training time in the air to reach a performance standard and First Shot measures, which look at initial levels of performance when the student transfers from simulator to aircraft. In flying training evaluations savings measures are more commonly used.

Expressing transfer in terms of the training time saved in the air through the introduction of a flight simulator yields useful information but it assumes that training in the flight simulator is not accountable. A more valid measure is what Roscoe (1980) has called the Transfer Effectiveness Ratio (TER).

$$\text{TER} = \frac{A - A_s}{S}$$

where A = aircraft training time when not using a simulator, A_s = aircraft training time when using a simulator, S = simulator training time.

A TER = + 1.0 will occur when the amount of training time saved in the air equals the amount of time spent in the simulator. TERs greater than + 1.0 arise when the amount of air time saved by the use of a simulator is greater than the time spent in the simulator and TERs less than + 1.0 result when more time is spent in the simulator than is saved in the air.

Orlansky & Chatelier (1983) reported the analysis of 22 studies of transfer of training using flight simulators conducted between 1967 and 1977. The median TER value for all these studies was 0.48 with a range from + 1.9 to − 0.4. A negative TER occurs when the introduction of a flight simulator results in more air time being required.

The literature on the use of the TER is extensive and the concept has been amplified and extended. It must suffice at this time to make a brief reference to the following developments:

(1) the value of the TER is enhanced by relating it to operating cost data obtained from the aircraft and the simulator. The TERs reported by Orlansky and Chatelier might appear disappointing as the majority fall below + 1.0. However, the Operating Cost Ratio (OCR) for simulator and aircraft is generally about 0.10. Consequently, a simulator producing a TER greater than + 0.10 can claim to be providing cost effective aircrew training. Note however that specific cost and training effectiveness information is needed in every case.

(2) the TER for a specific training task in the same simulator can change. A point can be arrived at where further use of the simulator may not produce a sufficient improvement in performance to warrant the additional expenditure of time and resources. To cope with this situation Roscoe (1971) formulated Incremental Transfer

Effectiveness. Povenmire and Roscoe (1973) applied the measure to the use of ground-based trainers at the Institute of Aviation at the University of Illinois. This technique combined with cumulative cost measures can provide valuable information with which to decide the extent of simulator employment. For a fuller examination of this point see Durose (1982)

(3)   the same simulator, used for a number of different training tasks, may generate a range of TERs. Applying the measure to specific elements of a training syllabus rather than global performance can provide an indication of those tasks for which the device is suited and those for which it is deficient. Holman (1979) made use of TERs to evaluate 24 syllabus elements in a military helicopter simulator training course. The TERs ranged from + 2.80 to 0.00. The low TERs showed those elements of the course where the visual system fitted to the device was unable to provide the cues needed to master the task

Not all training effectiveness decisions can be taken with the benefit of a transfer evaluation. For example with a new training system there may be no base line data available to provide a basis for comparison. In other situations, such as space, there may be no opportunity to train in the real environment. In such situations other evaluative designs are possible (Rolfe & Caro 1982).

11.11   General conclusions

(1)   the flight simulator can be an effective means of teaching, maintaining and assessing aircrew skills. To ensure effectiveness requires a detailed preliminary analysis of the training task, cooperation between customer, manufacturer, legislator and training technologist, during the preparation of a specification, and subsequent evaluation to verify training and cost effectiveness

(2)   substantial amounts of training time in the air can be replaced by the use of flight simulators. However, it should always be assumed that actual training time in the simulator will take at least as long as it would in the air. Any savings will accrue from the reductions in non-productive times, e.g., time spent transitting and positioning in order that a training detail can either commence or be repeated

(3)   the ease with which tasks learned in a flight simulator transfer to the aircraft varies. Procedural tasks can be learned more completely and transfer readily. Complex handling skills will demand greater levels of simulation in order to produce higher transfer

(4)   the effectiveness of simulation is not simply a function of the realism of the simulator. How the device is used and the quality of instruction will have a significant influence on learning

(5)   differences between the simulator and the actual equipment may be advantageous in order to improve the quality of training

(6)   motivation is a key element is achieving effective simulator training. Postive attitudes towards simulation must be held by instructors and communicated to the students

(7)   the simulator is one of a range of training devices which can supplement airborne experience. Every effort must be made to select and use the appropriate device

12

Résumé

12.1   Introduction

Most of the chapters of this book have been concerned with the techniques and technology required for the creation of a flight simulator, together with the physiological characteristics of people which that technology has to attempt to satisfy. We wish here to consolidate the essential features of these chapters, identifying the areas where the technology is likely to advance further and possible directions for that advance.

It is apparent that, in many parts of a flight simulator, the attempt to reproduce exactly the full details of the real world is bound to fail. Rather, the design of the simulator must be geared to the needs of the task and it is important to identify which elements of the simulation are required to satisfy these needs. The areas where exact duplication is, or may be, possible, such as the internal appearance of the cockpit or the flight and systems instruments, need not concern us here, even though we acknowledge that their simulation may not be easy. Neither need we consider other areas where duplication is either irrelevant, or even undesirable, such as the Instructor Operating Station. We concentrate on three areas where further elaboration of the inability to duplicate the aircraft may be useful and informative, viz.:

(1)   the computer system and its effect on the model
(2)   the motion system
(3)   the visual system

12.2   The computer system

Conventional aircraft are continuous systems. The movement of the pilot's controls (the inceptors), either by direct interconnection or via hydraulic or electrical actuators, moves the aircraft controls (the motivators) which in turn leads to response of the aircraft. Delays, if any, are due to lags in a continuous system; they do not contain discrete discontinuities. In early simulators, where the model was represented on an analogue computer, the dynamics of the aircraft could, in response terms, be perfectly reproduced. The computation was done in parallel using large numbers of computing elements, high gain amplifiers, each of which had a high bandwidth, of the order of Kilohertz. In human perceptual terms the computation was continuous, the dynamic response of the model was perfect, and no extra delay, compared with the aircraft was incurred. But the absolute accuracy of the analogue computer left something to be desired, its maintainability was a problem as its size increased to handle very large models of complicated aircraft, and its programability, particularly of non-linear functions, was difficult, its ability to store and replay information was limited, and so on.

The introduction of the digital computer solved these defects of the analogue computer but introduced a discrete computation cycle with a frame time, so that the computation proceeded at orders of tens of Hertz rather than of Kilohertz. Two prime consequences followed. Firstly the delay in response could become perceptible to the pilot, and this is especially so when compounded with additional delays from the visual system, as we shall see. Secondly, the accuracy of the response was degraded; for example, a poorly damped aircraft oscillation can become a divergent simulator oscillation.

There is therefore continuous pressure to increase the speed of computation. Individual computers are indeed increasing in speed, in part by the incorporation of several processors, but also by the utilisation of faster chips. The speed of light is rapidly becoming a limiting factor. The solution of a complex model in a reasonable frame time may only be possible by either subdividing the model into a number of sections running at different frame rates, determined by either modelling or perceptual needs, or by utilising a number of computers in parallel, or both. It is, in fact, now commonplace for simulators to use several mainframe computers with, at least in part, shared memory. Often, however, these multiple computers are separately dedicated to elements of the total simulation, such as the creation of the control feel forces, or part of the avionics or weapon system suite, with minimal interchange between computers, rather than representing true parallel computation.

Mainframe computers will continue to be used, particularly where data from the simulator requires to be collected and analysed in quantity – especially true of research and development simulators. But the introduction and widespread use in all walks of life of the microcomputer may lead us to suppose that the accent will be on parallel computers with a large number (of the order of dozens) of microprocessors solving the principal modelling equations and, in a sense, emulating the old analogue computation scenario. Many dozens more may be used for specific activities such as motion, instruments, control loading drives, as they are in part already. Currently, the vast majority of simulators have their shortest frame time set at 50 ms. Many research simulators operate at 20–30 ms and the aim here is to achieve 10 ms or less. While the ability to solve the model equations in this time is now reasonably routine, it is in the transfer of the solutions to and from the rest of the simulator where most benefit from the separate processors has been realised. By a strange quirk of fate, just as the ability to close the gap between the continuous process of the aircraft and the discrete nature of the digital computer is becoming a reality, so the aircraft, with the introduction of advanced control technology, is interposing a digital computer between the inceptors and the motivators. This computer modifies the pilot's demands according to laws designed to enhance the natural, unaugmented performance of the aircraft. However, the delays in response to pilot input introduced by this innovation led, in some early implementations, to quite unacceptable handling qualities. The analogy with some simulator problems was complete! Nevertheless, this trend will undoubtedly continue in aircraft so that the time is not far distant when the computation cycle of the simulator can in principle again correspond to that of the aircraft, at least in the representation of the more crucial dynamic responses. The increased power of the multiprocessor and associated high-speed computation will also assist in the stimulation of aircraft equipment, where this is a necessary simulation technique.

In spite of all these potential improvements the question of the task to be accomplished remains important. For the commercial, large transport simulator a short frame time may not be important – the aircraft modes are well damped and the response sluggish.

However, there is a great deal of computation to do if all the systems are to be faithfully represented. The light helicopter may, on the other hand, need appreciably less systems computation, though its flight model may be more complicated, but will have a demand for a short computation cycle in view of its agility and poorly damped dynamic modes. The balance between the two disparate requirements can lead to similar computation loads, the one with low repetition rates but large memory, the other with high repetition rate but less memory. The most demanding is, of course, the total simulation of an agile, military aircraft with a large number of complex systems and, in the battle damaged or emergency condition, very poor dynamics from the control point of view. It might therefore be deduced that this form of training simulator made the greatest computation demands, correctly so. But, in the training context, the problem is known, the aircraft exists. Even if the representation is not perfect it is possible, in principle, to determine if training is satisfactory. In the research simulator, either the aircraft does not exist, or the problem under investigation has not been studied in the particular form before. The demand for model accuracy is thus often overwhelming. The saving grace is that the study can often be conducted on a part task basis. It may be a handling qualities study so that the radio/navigation/weapons/hydraulic and electrical, etc., systems can be ignored, and all the computing power concentrated on the aircraft flight model, even ignoring undercarriage and ground effect if it is an up-and-away study. The availability of skilled test pilots, who are able to make a dispassionate analysis of the results of the simulation without being distracted by being launched into free air, is essential. Study of a mission simulation may also not require hydraulic and electrical systems and even may manage with a comparatively crude representation of the aircraft dynamics. These simplifications allow the computing power of the research simulator to be directed to where it is needed.

In summary then, there are now good prospects that a satisfactory calculation of the characteristics of the aircraft can be made. Perhaps the most difficult aspect of the operation, apart from the sheer programming effort, is the acquisition of accurate data.

12.3 Physiological stimulation

Kinaesthesis (motion and touch), vision, hearing are the essential senses stimulated in flight simulation. Smell and taste are generally ignored though the former may play some small part in the operation of aircraft. The information supplied to the auditory channel may be very representative of the real thing; it is not suggested that this is easy but it is, in principle, possible to provide noises which are practically indistinguishable from those present in the aircraft. It is not practical, however, to accomplish near duplication of motion and visual sensations, particularly in the representation of the real, visual, world. It is therefore important that we should analyse the stimulation of the senses provided by motion and vision to determine those features of the total environment which are important.

A starting point is to endeavour to identify what information is received by the human sensors and then to model the manner in which this information is interpreted. What is utilised, and how, will clearly be task dependent. It would be convenient if everybody chose the same information and interpreted it in the same way for a given situation. There is in fact considerable evidence that the interpretation of a given set of limited information varies widely between individuals (e.g. Palmer & Petitt 1976). And it is limited information with which we are concerned. We are born with certain in-built routines in the brain to allow basic survival, but with a large empty space waiting for a program. The manner in which this program develops depends, in both the broad and narrow sense, on the experience of the individual. While it might be hoped that the decision to become a pilot, for example, was influenced by a predisposition to view the world in a certain way, it clearly will not represent general uniformity across all elements of the piloting task, and certainly not to the relatively limited elements of the perception of motion and visual inputs and their interpretation. We are therefore faced with a potentially insuperable dilemma. We can produce only a very limited subset of the information available in the real world but this subset, while potentially useful to some individuals, may not be appropriate to others.

Fortunately the human being is very adaptable, and the world is full of redundant information. So if preferred information is absent then some other relevant, though perhaps less easily interpreted, information will be selected. The limits of operation of aircraft are often determined by the reduction of information in the real world to a minimum, and the substitution of artificial aids. For example, the lighting patterns and runway markings of airfields are designed to allow aircraft to approach and land in conditions where information from the natural world not only ceases to be redundant but is totally inadequate, due to poor visibility or darkness. It is perhaps not surprising that it is these conditions that modern visual systems of simulators are best able to reproduce.

Armed with some knowledge of the intrinsic capability of human sensors and the adaptability of people in making use of the available data or cues, whether by past experience or specific training, we are able to identify some general rules for utilising the capabilities of the technology. The following sections relate these to the motion system and the visual system as a summary of the more extensive treatment of Chapters 6 and 7.

12.3.1   *The motion system*

The obvious inability to repeat, on the ground, the true motion of an aircraft in all phases of flight means that the manner of operation depends upon the perceptual capabilities of the human being. The modelling of the human sensors is now well advanced (Borah *et al.* 1979) so that the conscious perception of motion can be predicted and the appropriate drive algorithms created. While there is some evidence (Nashner 1970) that unconscious perception may influence behaviour, and the models of the perceptual processes are not well conditioned to take account of this, nevertheless, far more important is the identification of conditions under which, in spite of an acknowledged influence of motion on behaviour, there may be no important effect on the performance of the task being simulated. This is particularly so in the training process. The US Air Force, for example, decided in 1980 that military training simulators would not have motion systems as there was no reliable evidence that motion affected training transfer. The Federal Aviation Administration, on the other hand, in both Phase II and Phase III requirements (FAA 1983) insists on motion systems for the training simulators of the operators of commercial transports. Since military aircraft may be highly manoeuvrable, suffer stability deficiencies in failure conditions, have tasks which demand tight, closed loop, control during tracking, all conditions where the influence of motion on pilot behaviour is most pronounced, whereas commercial aircraft generally fall into none of these categories, these decisions as to when motion is and is not necessary could be considered perverse. They indicate, however, that the training process can proceed without replication of the trainee's actions in the real world.

So, in spite of the understanding of the influence of motion on pilot behaviour, the conditions under which it is needed are not well determined. Two features are, however, universally agreed as vital and, because of a failure to satisfy these features, many studies of the part played by the motion system in the performance of a simulated task are of doubtful validity. The first is the importance of the quality of the motion system, the mechanics of producing the motion, so that no detectable unwanted inputs to the pilot in the form of bumps and jerks – mechanical noise – are generated. The second is the quality of the drive algorithm, the phasing of the command input, the strength of the washout, so that the spurious motions which generally are present are not sensed. The ability to accomplish both of these aims (undetectable noise and imperceptible washout) is totally dependent on the fact that man has a threshold of conscious perception and the presumption that any input below this threshold will have no influence on his response.

More contentiously, it can be argued that because the motion cue provided will be of shorter duration than that in reality, and may be of smaller initial magnitude, the pilot will have to go through a process of adaptation, and as a consequence it may be important that the same cue should always be given for the same aircraft response. The contrary argument would contend that it is not possible to adapt to short term motion cues, our responses are 'hard wired' and so the closest approximation to the real cue should always be given even though this may lead to different cues about a given axis as a function of aircraft manoeuvre. The argument is relevant particularly to synergistic systems where the maximum available acceleration about an axis, in magnitude possibly and in duration certainly, is dependent on the displacements at the time along and about other axes. On the other hand the thresholds of perception of aircrew, and those who fly very frequently, are known to be elevated with respect to those of the general population; some element of adaptation therefore does take place, albeit over an extended period of time.

Chapter 6 indicates clearly that much is now known about the significance of the motion cue and that the identification of suitable drive algorithms is well advanced. In other words the motion systems of simulators serve their purpose in most instances. Nevertheless there is still dispute as to the correct manner of driving the system (Parrish *et al.* 1975, 1976) and about the only firmly established fact is that spurious inputs to the pilot exceeding his perceptual thresholds, be they from mechanical noise or washout, will lead to an unacceptable system, eliciting incorrect responses in the research simulator and having a possible deleterious effect on transfer of training.

Most of the studies of motion have been directed at systems which move the whole cockpit. The obvious inability of conventional systems to simulate sustained normal acceleration led to the consideration of 'g'-seats', where pressures on the tuberosities were meant to compensate for the non-existent stimulation of the otoliths. But 'g'-seats', enhanced if need be with actuators to pitch, roll and yaw them, are also capable of providing onset cues. There may result an incompatibility in pilot location with respect to the rest of the cockpit but, in the military context, this already exists in extreme manoeuvres with conventional motion systems; a 'g'-seat' can be arranged to actually improve the representation under these conditions. Several studies of 'g'-seats' have been conducted, some with quite favourable results and others less so (Ashworth, McKissick & Parrish 1984; Parrish & Steinmetz 1983; Showalter & Parris 1980), but the attraction of dispensing with the motion system is sufficient to warrant further development of this technique of motion cuing. This is particularly so with respect to some solutions to the visual display needs, which may be incompatible with a motion system.

Chapter 7 describes the concept of Area of Interest visual systems, be they for combat simulators or to cover more general tasks. While it would be premature to identify which of the alternative display techniques for AOI systems will be the most generally used, a powerful contender in some circumstances, and currently most widespread, is projection onto the inside of a comparatively large dome. This seriously degrades the performance of any motion system on which it is mounted and may totally inhibit the use of a motion system. The attraction of the 'g'-seat' under these circumstances is clear. Following further research and development their use in conjunction with a 'g'-suit' for the simulation of sustained 'g' may well prevail.

12.3.2 *Visual systems*

An oft quoted statistic is that a decent sized oak tree has some 250 000 separately identifiable polygons. The most advanced, and expensive, visual systems currently offer some 5000 polygons for viewing of a complete scene at any one time, and are striving for perhaps 50 000 in the future. There are thus difficulties in attempting to duplicate the real world. While the polygons of an oak tree may easily be significant to the pilot of a hovering helicopter waiting in ambush, all 250 000 can not be observed simultaneously by one individual. Thus, where moderately faithful representation of the real world is necessary, and this is by no means always so, presenting the information only where the observer is looking reduces the polygon requirement considerably. Hence the interest in Area of Interest displays, particularly those that are eye slaved.

The benefits are twofold. The smaller field of view encompassed by the display means that, for a given bandwidth of the display device, a higher resolution of the image is achievable and, without extreme demands on the technology, it is possible to approach the resolving power of the eye. Secondly, all the available scene content can be compressed into a smaller area, thus enriching the picture. It should be appreciated, however, that this second benefit only represents, at most, a six-fold (generally nearer four-fold) increase in possible polygon density, for a single channel. In one sense this is as well as the cost of producing the data base for a high scene content is already substantial; increasing it by orders of magnitude could be untenable.

The availability of digital data bases, Digital Mapping Agency data, provides a potential relief in that a capability for automatic generation of the CGI data base has been achieved. The cultural data within the DMA data bases is inadequate for the representation of a particular area in detail so that an element of manual, hand-crafting, remains for some applications. This lack of detail will also encourage the incorporation of texture into the capabilities of the image generator, which effectively provides large numbers of apparent surfaces without in fact consuming any from the generator itself. The consequent satisfying scene richness also contains important cues as to range from features and height above the ground.

Nevertheless, we should not overlook the inherent capability of some display techniques to provide a three-dimensional image. An example is the helmet mounted display being developed by CAE, in Montreal, Canada.

Fig. 12.1 shows how a separate image is presented to each eye and in this AOI display the high resolution area is totally binocular, there being about a 30° common area, viewed by both eyes, in the middle of the total 135° azimuth field of view of the background, low resolution, imagery.

Of course the question of adequate scene content is intimately related to the ability to process the imagery quickly. There are two aspects of this 'real-time' requirement. One is the need to pass quickly down the pipeline of the image generator process so that the picture ultimately presented to the pilot arrives at a time not too long delayed after the events which it represents. In other words, to reduce what is called the transport lag to a minimum. The other is the need to update the scene content sufficiently frequently to ensure that objects appear to move smoothly. The second clearly contributes to the first; but both mean that the time that can be

Fig. 12.1. Helmet-mounted Area of Interest display.

spent on processing successive scenes is very limited compared, for example, with advertising television films, which can be created and recorded frame-by-frame at any convenient rate and then run through at the correct speed. The magnitude of the possible total delay, taking into account the frame time of the host computer of the simulator, should not be underestimated. Some CGI systems update at TV frame rate (every 40 ms in UK) though most do so at field rate. If we take a typical frame time of 50 ms for the host computer, a computation time for the CGI of two frames, then we arrive, in the worst case, at a delay between a step input to the controls and the presentation of the relevant picture as shown in Fig. 12.2. The various inputs just miss the start of a frame. This system has a longish but not unreasonable transport lag of 100 ms, a quite usual frame time for the host computer and yet illustrates a 240 ms interval between control input and completion of the first display field. Admittedly a worst case but an average delay would be 195 ms. Except perhaps for the most sluggish of aircraft these delays are unacceptable. A substantial improvement can be made by matching the frame time of the host computer to that of the CGI (i.e., 40 ms) and arranging for synchronous operation. The worst case then drops to 160 ms, with an average of 140 ms, which is still rather long.

To illustrate the importance of frame time, from whatever the source, we have deliberately chosen an unreal example. Practically, the technique is to use derivative (velocity, perhaps acceleration) data as well as position to predict ahead. If the aircraft motion could be represented by a stationery process the picture presented would then be accurate both in time and location. For common aircraft motions the question actually becomes one of the accuracy of the position depicted rather than a straightforward delay.

From the above discussion and the details of Chapter 7 it is evident that the visual systems of simulators are in a state of rampant development, but that there is still a long way to go before all potential users are satisfied.

Fig. 12.2. Interval between control input and scene display.

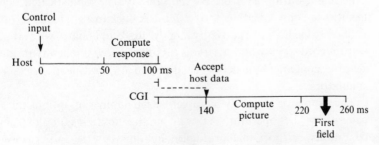

While the way ahead is by no means blocked, the technology has not reached a dead-end, a break-through into some new innovative technique would still be welcome.

12.4   Other challenges

We pick up here a few aspects of the current state of the art where progress may be rapid and close with a sample of the disciplines appropriate to flight simulation.

12.4.1   *Instructor operating stations*

This vital part of the simulator is one of the few which is not a simulation. It is also a part that can be the subject of much criticism by users. What may be known as the Control Desk or Operator's Console of the research and development simulator is similarly important and potentially equally irritating. The difficulty is that in these areas the set-up, start-up, shut-down and general control of the progress of the simulation occurs with the operator's intervention, whereas he actually wants to instruct a student, or perform, say, a handling qualities or avionics experiment. Admittedly, in the former case, the instructor may wish only to interact with the student, whereas in the latter, the operator may wish to interact with the 'model' as well as the pilot. There is thus a fine balance between the automation of the sortie so that the operator can pursue his prime task, and the retention of flexibility to allow an on-line change of plan in the light of circumstances during the sortie.

One of the main problems is that the need to interact with the simulator tends to occur at the same time as the need to communicate with, or monitor the actions of, the student/pilot. The operator is thus likely to miss responses which are the main purpose of the sortie. This is clearly important in the training environment and there is evidence that instructors do miss a great many student errors (Thanderz 1982) but even in the research simulator the gathering of data and subsequent analysis is not sufficient without pilot interaction, which indeed may alter the course of an experiment during a sortie. By the same token, sophisticated debriefing facilities are only an adjunct to on-line feedback to a student. Nevertheless, debriefing facilities are currently largely 'chalk and board' with straightforward record–replay and a serious attempt is now underway to identify a sensible content for a debrief facility and its implementation on future simulators.

The simulator generally has a very large number of options that can be chosen by the operator. As well as the aircraft and environmental conditions – airport, meteorological, lighting patterns, targets, etc. – there are also large numbers of selectable failure modes. All of these appear on many pages of a VDU, which will also contain feedback of the progress of the flight (digital data and soft instruments). While the flight simulator is itself a clear case of computer based training there is a strong case for computer based monitoring and control of the simulator and the use of intelligent knowledge based systems (IKBS) at the instructor station. An IKBS would be able to manage the selection of displayed pages, transmit unrequested but relevant information on the basis, initially, of predefined objectives of the simulator exercise and later, on the experience gained during multiple similar exercises, and assist in the analysis of the student responses according to training criteria. A further stage would be to employ the same system to guide and implement the debriefing information to be displayed to the student after the sortie. Clearly, care will be needed to ensure that such a system does not make it more difficult to intervene and redirect the sortie according to the requirements of the instructor rather than those of the IKBS.

12.4.2 *Complex aircraft systems*

Because the training simulator is now such a vital part of any aircraft operation it is slowly being recognised as an essential element in the total equipment required to fulfill the operational needs. Appreciating this, some aircraft systems, because of their complexity need to have incorporated within their design a capability for being inserted into, and stimulated by, the simulator. Primarily this arises from the complexity of the software and the possibility of revisions during the life of the simulator. The consequences for the aircraft equipment are that it must be capable of being updated at the simulator rate and it must accommodate the compute, freeze, reset functions of the simulator. The freeze facility, in part or whole, is one of the most frequently used instructor tools and must be useable. It normally requires special provision.

12.4.3 *Applications*

Chapters 10 and 11 departed from the theme of the elements of flight simulation to consider its application in the areas of research and training. Both fields present their own unique challenges and rewards. For the design engineer the simulator can provide data which will help in the development and evaluation of concepts and options. For the flight dynamicist it is a way of examining the behaviour of the vehicle under flight conditions and to suggest to the aerodynamicist potential ways of improvement. To the human factors specialist it is a means of optimising the allocation of functions between man and machine, the control display interface and operating environment. The psychologist and physiologist have in the simulator a medium in which to measure operator performance capabilities and responses to the stresses encountered in flight. For the instructor it is a device that allows training to be undertaken safely and effectively. To the flight operations examiner it provides a vehicle in which to assess competence.

Consideration of the applications of flight simulation has not been the primary purpose of this book; indeed the topic is one that warrants another volume similar in length to the present text. However, in both the research and training applications of flight simulation the message is a clear one. The simulator can be an indispensible means of achieving the task. But, it must always be remembered that simulation is only a tool and as such its effectiveness depends on the ability of the designer to identify the essential elements needed in the simulation and the ability of the user to employ the device in the correct manner. Fraser (1966) put forward the following principles as a warning to those who might naïvely consider that simulation-based research might provide a solution to most of their problems:

(1)   a situation, task, or problem can be simulated in a valid manner only insofar as its parameters are known

(2)   a simulation can provide the solution to a problem or provide valid information, only it the elements of the solution or the information reside in that simulation

(3)   simulation may provide an incorrect solution or false information if the simulation is incomplete or the parameters of the simulation are incorrect

(4)   only those parameters of a situation, task, or problem necessary for the completeness of a simulation need be represented in the simulation

In a complementary fashion Morgan (1971) offered the following caution with regard to the exploitation of computer-based training systems, of which the flight simulator is one example:

(1)   don't act as if computer-based training is something entirely new. Always conceive of it in such a way that it can be related to what is known about conditions for effective training

(2)   don't be misled into thinking that a simulator offers a training system which is obviously so superior that no evaluation need be made or no records kept

(3)   don't assume that a simulator can simply be plugged into the training programme. Carefully analyse the total training process

and apply the simulator to those functions for which it is best suited

These comments emphasise the importance of the simulator designer and potential user working closely with one another. Each has different objectives; the one, given the opportunity, to design the ultimate simulation; the other to obtain the simulation that best fits the requirements of the application. Neither objective is misguided or wrong but each must be seen for what it is and the situation of a potential conflict recognised.

12.5   Concluding remarks

The editors of this book set themselves the task of seeking out authors who, together with themselves, would provide a description of the theory and practice which makes up the field of technology associated with the design, construction and operation of flight simulators. In attempting this task the chapters in the book have identified the principal elements making up the flight simulator. In so doing it will be seen that simulation not only applies but encourages new technology and that this application of new ideas ranges widely from the aeronautical to the behavioural sciences. It encourages the creativity of the mathematician and creative graphic artist. It brings all the contributions together to produce dynamic illusions that are so real that astronauts are familiar with the moon before their first visit and airline pilots can acquire the skills of captaincy without leaving the ground. However, achievements of this stature are arrived at not by persuading the user to suspend disbelief but by producing a simulation that can pass the most critical examination.

What of the future? Certainly advances in new technology will be used to enhance simulation capability. At the same time the nature of the simulation is likely to change. The growing power of the small computer, coupled with new head-mounted display devices and dynamic seats, may make the large flight simulators of today look as cumbersome as the dinosaur appears now against the pattern of evolutionary development. The simulator as an independent device may change to become an imbedded element in the actual operating system. On another dimension the application of simulation for the purpose of pleasure, referred to in the first chapter, also offers some interesting challenges. The advent of 'theme parks' like Disneyland at one level and the market for flight simulation software for the home micro-computer at the other, are examples of the range of demand that presents itself. The flight simulator manufacturers have been quick to realise the potential entertainment value of the knowledge that they possess. For example, Rediffusion's first leisure application, a 40-seat space ship simulator with visual and motion cues, was operational in Toronto in 1985.

From humble beginnings to the modern multi-million pound facility the flight simulator has made dramatic advances and contributed significantly to the improvement of aircraft design, operations and safety. Challenges remain and will continue to do so, so long as there are new aerospace vehicles to create and new generations of crews to train. The editors hope that this text has encapsulated a little of the knowledge that currently exists and has presented it in such a way as to help make the future of flight simulation more sound and secure.

References

Chapter 1

Abt, C. C. (1964). War gaming. *International Science & Technology*, **32**, 29–37.
Davis, D. P. (1975). Approval of flight simulator flying qualities. *The Aeronautical Journal*, **79**, 281–98.
Elmaghraby, S. E. (1968). The role of modelling in IE design. *Journal of Industrial Engineering*, **19**, No. 6.
Harward, D. M. (1910). The Sanders Teacher. *Flight*, **2**(50) 10 December, 1006–7.
Kelly, L. L. & Parke, B. P. (1970). *The Pilot Marker*. New York: Grosset & Dunlap.
Prophet, W. W., Caro, P. W. & Hall, E. R. (1971). Some current issues in the design of flight training devices. In *25th Anniversary Commemorative Technical Journal*, US Naval Training Devices Center, Orlando, Florida.
Reid, G. H. & Burton, H. L. (1924). Psychomotor responses in relation to flying. *Proceedings of the Royal Society of Medicine*, **17**, 43–53.
Scans, N. S. & Barnes, A. G. (1979). Fifty years of success and failure in flight simulation. In *Fifty Years of Flight Simulation*. London: Royal Aeronautical Society.
Shannon, R. E. (1975). *Systems Simulation; The Art and Science*. Englewood Cliffs: Prentice-Hall.
Tabs (1964). *Golden Jubilee Number*, vol. 22, 1, 82–5. London: Strand Electrical Co. Ltd.

Chapter 2

Adorian, P., Staynes, W. & Bolton, M. (1979). The evolution of the flight simulator. In '*Fifty Years of Flight Simulation*'. London: Royal Aeronautical Society.
Air Member for Training (1945). *Notes on the History of RAF Training 1939–1944*. London: Air Ministry.
Anderson, H. G. (1919). *The Medical and Surgical Aspects of Aviation*. Henry Froude, Oxford University Press.
Bairstow, L. (1920). *Applied Aerodynamics*. London: Longman, Green & Co.
Bennett, S. (1979). *A History of Control Engineering 1800–1930*. Stevenage: Peter Peregrinus.
Billing, E. (1910). An improved machine for teaching the art of flying without leaving the ground. British Patent Specification 16 773.
Brice, D. A. (1951). A pilot looks at the Dehmel. *The Aeroplane*, January, **19**, 73–6.
Clark, C. C. (1962). Human control performance and tolerance under severe complex wave form vibration with a preliminary historical review of flight simulation.

Proceedings of the National Meeting on Manned Space Flight. St Louis, Missouri, April 30–May 2, 176–94.

Curtis, A. H. (1978). Synthetic training origins in the RAF. *Air Pictorial,* **40**(4), 151–2.

Cutler, A. E. (1966). Environmental realism in flight simulators. *The Radio and Electronic Engineer,* January, 1–12.

de Florez, L. (1949). Synthetic aircraft. *Aeronautical Engineering Review,* April, 26–9.

Dehmel, R. C. (1941). US Patent Specifications 2, 494, 508, 2, 366, 603.

Deighton, L. (1977). *Fighter.* London: Jonathan Cape.

Directorate of Operational Training. (1942). *Illustrated Catalogue of Synthetic Training Devices.* London: Air Ministry.

Dreves, R. G., Pomeroy, R. R. & Voss, H. A. (1971). The history of training devices at the Naval Training Device Center. In *25th Anniversary Commemorative Technical Journal,* pp. 1–10. US Naval Training Device Center, Orlando, Florida.

Dummer, G. W. A. (1985). The story of radar training devices. *Electronics & Power,* June, 455–460.

Elson, B. M. (1967). Color TV generated by computer to evaluate spaceborne systems. *Aviation Week and Space and Technology,* 30 October, pp. 78, 81, 82.

Flight Training Research Association. (1940). British Patent 577, 566.

Haward, D. M. (1910). The Sanders Teacher, *Flight,* **2**(50), 10 December, 1006–7.

Hill, N. & de Guillenschmidt, P. (1955). The possibility of the flight simulator as a training aid to helicopter pilots. *The Journal of the Helicopter Association of Great Britain,* **8**(4), 133–57.

Hooven, F. J. (1978). The Wright brother's flight-control system. *Scientific American,* **239**(5), 132–40.

Huff, T. (1980). Developing the flight crew simulator. *Naval Aviation News,* February, 14–19.

Janes Aircraft. (1919), p. 386, article on Gabardini.

Jenkins, H. F. & Berlyn, R. C. (1932). British Patent Specifications 396, 537/38/39/40.

Johnson, W. E. P. (1931). Apparatus for use in flying training. British Patent Specification 378, 172.

Kelly, L. L., as told to Parke, R. B. (1970). *The Pilot Maker.* New York: Grosset & Dunlap.

Lender, M. & Heidelberg, P. (1917). Improvements relating to apparatus for training aviators. British Patent Specification 158, 522.

Lender, M. & Heidelberg, P. (1981). Safe apparatus for instruction in the management of military aeroplanes. British Patent Specification 127, 820.

Link, E. A., Jr. (1930). Combination training device for student aviators and student entertainment apparatus. US Patent Specification 1, 825, 462.

Maisel, A. Q. (1944). They bombed Berlin in Binghamton. *Reader's Digest,* April, 45–6.

Mueller, R. K. (1936). An electrical device for solving the equations of longitudinal stability. *Journal of the Aeronautical Sciences,* **3**, March, 158–60.

Murray, F. J. (1971). Training devices computing systems, In *25th Anniversary Commemorative Technical Journal,* pp. 221–30. US Naval Training Device Center, Orlando, Florida.

Redmond, K. C & Smith, T. M. (1980). *Project Whirlwind The History of a Pioneer Computer.* Bedford, MA: Digital Press.

Reid, G. H. & Burton, H. L. (1924). Psychomotor responses in relation to flying. *Proceedings of the Royal Society of Medicine,* **17**, 43–53.

Robinson, D. H. (1973). *The Dangerous Sky; A History of Aviation Medicine.* Henley-on-Thames: G. T. Foulis & Co.

Roeder, H. A. (1929). Ubungsgerat zum Erlernen der Steurung eines im Raum frei beweglichen Fahrzeuges. German Patent Specification 568, 731.

Rougerie, L. (1928). Banc d'entrainement au pilotage sans visibilite exterieure. French Patent Specification 655, 874.

Ruggles, W. G. (1917). Orientator. US Patent Specification 1, 342, 871.

Ruggles, W. G. (1918). Orientator. US Patent Specification 1, 393, 456.

Soroka, W. W. (1954). *Analog Methods in Computation and Simulation*. New York: McGraw-Hill.

Taylor, J. W. R. (1958). *C. F. S. Birthplace of Air Power*. London: Putnam.

Walters, W. H. (1910). Apparatus for teaching the art of aeroplaning, applicable also for amusement purposes, British Patent Specification 9, 950.

Williams, F. C. & Uttley, A. M. (1946). The velodyne. *Journal of the Institute of Electrical Engineers*, **93**(IIIA), 1256–74.

Winslow, Carol Dana. (1917). *With the French Flying Corps*. London: Constable & Co.

Winter, D. (1982). *The First of the Few*. London: Allen Lane.

Chapter 3

Babister, A. W. (1961). *Aircraft Stability and Control*. Oxford: Pergamon Press.

Bramwell, A. R. S. (1976). *Helicopter Dynamics*. London: Edward Arnold.

ESDU (1966). The equations of motion of a rigid aircraft. Item 67003, *Engineering Sciences Data, Dynamics Sub-Series*. London: Engineering Sciences Data Unit.

ESDU (1967a). Conversion formulae for rotation and translation of axes. Item 67004, *Engineering Sciences Data, Dynamics Sub-Series*. London: Engineering Sciences Data Unit.

ESDU (1967b). Geometric and kinematic relationships for various axis systems. Item 67036, *Engineering Sciences Data, Dynamics Sub-Series*. London: Engineering Sciences Data Unit.

ESDU (1967c). Direction and incidence angles. Item 67037, *Engineering Sciences Data, Dynamics Sub-Series*. London: Engineering Sciences Data Unit.

ESDU (1968). Atmospheric data for performance calculations. Item 68046, *Engineering Sciences Data, Dynamic Sub-Series*. London: Engineering Sciences Data Unit.

Etkin, B. (1959). *Dynamics of Flight*. New York: John Wiley.

Fang, Ai Chun & Zimmerman, B. G. (1969). *Digital Simulation of Rotational Kinematics*. NASA TN D-5302.

Hanke, C. R. & Nordwall, D. R. (1971). *The Simulation of a large Jet Transport Aircraft*. Vol. 1, *Mathematical Model*, Vol. 2, *Modelling Data* NASA CR 1756.

Hopkin, H. R. (1966). *A Scheme of Notation and Nomenclature for Aircraft Dynamics and Associated Aerodynamics*. RAE Technical Report TR 66200, Parts 1–5 (also ARC R and M 3562, Parts 1–5).

Jansen, C. J. (1982). Non-Gaussian atmospheric turbulence model for flight simulator research. *Journal of Aircraft*, **19**, 5, 374–9. (Also AIAA Paper 80–1568.)

Klehr, J. T. (1983). Meteorological inputs for flight simulators. *Journal of Aircraft* **20**, No. 1, 91–2. Also AIAA Paper 82–0216.

Klehr, J. T. (1984). The simulation of hazardous flight conditions. AIAA paper 84–0278.

McFarland, R. E. (1975). *A Standard Kinematic Model for Flight Simulation at NASA–Ames*, NASA CR 2497.

Mitchell, E. E. L. & Rogers, A. E. (1965). Quaternion parameters in the simulation of a spinning rigid body. *Simulation*, **18**, No. 6.

Robinson, A. C. (1958). *On the use of Quaternions in the Simulation of Rigid Body Motion*. WADC TR 58-17.

Schlesinger, S. (1979). Terminology for model credibility. *Simulation*, **32**, 3 103–4.

Thomas, H. H. B. M. (1984). Some thoughts on mathematical models for flight dynamics. Aeronautical Journal **88**, No. 875, 169–78.

Tomlinson, B. N. (1975). Developments in the simulation of atmospheric turbulence. In *Flight Simulation/Guidance Systems Simulation*. AGARD CP-198. Paris, Advisory Group on Aerospace Research and Development.

Tomlinson, B. N. (1979). *SESAME—A System of Equations for the Simulation of Aircraft in a Modular Environment*. RAE Technical Report 79008.

Chapter 5

ARINC (1981). *Mark 33 Digital Information Transfer System (DITS)*. ARINC 429–5–81, Washington D.C., Aeronautical Radio Inc.

CAA *Approval of Simulators for the Training and Testing of Flight and Cabin Crews in Civil Aviation*. UK Civil Aviation Authority, CAP 453 (to be published).

Cartwright, P. A. & Wearne, S. H. (1984). *Health and Safety Legislation and Engineers*, Proceedings of the Institute of Mechnical Engineers, **198B** No. 10.

FAA (1983). *Airplane Simulator and Visual System Evaluation*, US Federal Aviation Administration. Advisory Circular 120–40, 1/31/83.

FCC (1980). Title 4 7, *Code of Federal Regulations, Telecommunication*, Chapter 1. Federal Communication Commission, Part 15, US Government Printing Office.

MIL STD 1553B, (1980). *Aircraft Internal Time Division Command/Response Multiplex Data Bus*, Washington D.C., Department of Defense.

NFPA (1984). *Standard on Halogenated Fire Extinguishing Agent Systems – Halon* 1301, Boston, National Fire Protection Association, No. 12 A.

Chapter 6

AGARD (1979). *Dynamic Characteristics of Flight Simulator Motion Systems*, AGARD-AR-144. Paris, Advisory Group on Aerospace Research and Development.

AGARD, (1980). *Fidelity of Simulation for Pilot Training*, AGARD AR-159. Paris, Advisory Group on Aerospace Research and Development.

Albery, W. B. (1981). Development of a Flight Simulation Capability in the Dynamic Environment Simulator. AIAA paper 81–0978. In *AIAA Flight Simulation Technologies Conference, a Collection of Technical Papers*, pp. 92–7. New York: American Institute of Aeronautics and Astronautics.

Ashworth, B. R. (1976). *A Seat Cushion to Provide Realistic Acceleration Cues for Aircraft Simulators*. NASA TM X-73945. Washington: National Aeronautics and Space Administration.

Ashworth, B. R. & McKissick, B. T. (1978). The effect of helmet loader G-cueing on pilot's simulator performance. In *Flight Simulation Technologies Conference, Arlington TX, 18–20 September, Technical Papers*. pp. 15–20. New York: American Institute of Aeronautics and Astronautics.

Baret, M. (1978). Six degrees of freedom in large motion system for flight simulation. In *Piloted Aircraft Simulation Techniques*. AGARD-CP-249, pp. 22–1 – 22–8. Paris, Advisory Group on Aerospace Research and Development.

Baron, S., Lancraft, R. & Zacharias, G. (1980). *Pilot/Vehicle Model Analysis of Visual and Motion Cue Requirements in Flight Simulation*. NASA CR 3312. Washington: National Aeronautics and Space Administration.

Baron, S., Muralidharan, R. & Kleinman, D. (1980). *Closed Loop Models for Analyzing Engineering Requirements for Simulators*. NASA CR 2965. Washington: National Aeronautics and Space Administration.

Baron, S. (1983a). An Optimal Control Model Analysis of Data from a Simulated Hover Task. In *Eighteenth Annual Conference on Manual Control, Dayton, OH, 8–10 June*, pp. 186–206. AFWAL-TR-83-3021. Dayton, Ohio: Air Force Wright Aeronautical Laboratories.

Baron, S. (1983b). *A Pilot/Vehicle Model Analysis of the Effects of Motion Cues on Harrier Control Tasks*. NAVTRAEQUIPCEN 80-D-0014-0019-1. Orlando FL: Naval Training Equipment Center.

Borah, J., Young, L. R. & Curry, R. E. (1979). *Sensory Mechanism Modelling* AFHRL-TR-78-83. Dayton, Ohio: Human Resources Laboratory.

Bose, E. B., Leavy, W. P. & Ramachandran, S. (1981). Improved G-cueing system. AIAA paper no. 81–0987. In *AIAA Flight Simulation Technologies Conference, a Collection of Technical Papers*, pp. 139–46. New York: American Institute of Aeronautics and Astronautics.

Bray, R. S. (1972). *Initial Operating Experience with an Aircraft Simulator Having Extensive Lateral Motion*. NASA TM X-62, 155. Washington: National Aeronautics and Space Administration. [T85]

Bray, R. S. (1982). Helicopter simulation technology: an Ames Research Center perspective. In *Proceedings of a Specialists Meeting on Helicopter Handling Qualities, NASA Ames Research Center, Moffett Field, CA, U.S.A., 14–15 April*, NASA CP 2219, pp. 199–208. Washington: National Aeronautics and Space Administration.

Cardullo, F. M. & Kosut, R. L. (1979). A systems approach to the perception of motion in flight simulation. In *50 Years of Flight Simulation*, London: Royal Aeronautical Society.

Cardullo, F. M. (1981). Physiological effects of high-G flight: their impact on simulator design. AIAA paper no. 81–0986. In *AIAA Flight Simulation Technologies Conference, a Collection of Technial Papers*, pp. 147–53. New York: American Institute of Aeronautics and Astronautics.

Caro, P. W. (1977). Platform motion and simulator training effectiveness. In *Proceedings of the 10th NTEC/Industry Conference, Orlando, FL, 15–17 November*, NAVTRAEQUIPCEN IH-294, pp. 93–7.

Caro, P. W. (1979). The relationship between flight simulator motion and training requirements. *Human Factors* 21(4), pp. 493–501.

Cooper, G. E. & Drinkwater, F. J., III (1971). Pilot assesment aspects of simulation. In *Conference Proceedings No. 79, Simulation*, pp. 9–1 – 9–18. AGARD-CP-79-70. Neuilly sur Seine: Advisory Group on Aerospace Research and Development.

Crosbie, R. J. (1983). Application of experimentally derived perceptual angular response transfer functions. AIAA paper no. 83–1100. In *1983 Simulation Technologies Conference, Collection of Technical Papers*, pp. 146–53. New York: American Institute of Aeronautics and Astronautics. [T33]

Crosbie, R. J. & Elyth, J. Jr. (1983). *A Total G-Force Environment Dynamic Flight Simulator – A New Dimension in Flight Simulation*. AIAA paper no. 83–0139.

DeBerg, O. K., McFarland, B. P. & Showalter, T. W. (1976). The effect of simulator fidelity on engine failure training in the KC-135 Aircraft. In *Proceedings of the AIAA Visual and Motion Simulation Conference, Dayton, Ohio, 26–28 April*, pp. 83–7. New York: American Institute of Aeronautics and Astronautics. [T33]

Dieudonne, J. E., Parrish, R. V. & Bardusch, R. E. (1972). *An Actuator Extension Transformation for a Motion Simulator and an Inverse Transformation Applying Newton-Raphson's Method*. NASA TN D-7067. Washington: National Aeronautics and Space Administration.

Douvillier, J. G., Turner, H. L., McLean, J. D. & Heinle, D. R. (1960) *Effects of Flight Simulator Motion on Pilots' Performance in a Tracking Task*. NASA-TN-D-143. Washington: National Aeronautics and Space Administration.

Dusterberry, J. C. (1974). Potential use of engineering simulators to increase training economy. *Training Economy Through Simulation, NTEC/Industry Conference Proceedings, Orlando, Florida, 19–21 November*, pp. 163–6, NAVTRAEQUIPCEN IH-240. Orlando: Naval Training Equipment Center.

Dusterberry, J. C. & White, M. D. (1979). Development and use of large-motion simulator systems in aeronautical research and development. In *50 Years of Flight Stimulation*. London: Royal Aeronautical Society.

Engelbert, D. F., Bakke, A. P., Chargin, M. K. & Vallotton, W. C. (1976). Mechanical design of NASA Ames Research Center Vertical Motion Simulator. In *Tenth Aerospace Mechanisms Symposium*, pp. 155–64. NASA CR-148515. Washington: National Aeronautics and Space Administration.

FAA (1980). *Aircraft Simulator and Visual System Evaluation and Approval*, AC 121–14C, Washington: Department of Transportation, Federal Aviation Administration.

Gallagher, J. T. (1970). Requirements for simulators used in handling qualities research. AIAA paper no. 70–353. In *Proceedings of AIAA Visual and Motion Simulation Technology Conference, Cape Canaveral FL, 16–18 March*, AIAA paper no. 70–353. New York: American Institute of Aeronautics and Astronautics.

Gilbert, W. P. & Nguyen, L. T. (1978). Use of piloted simulation for studies of fighter departure/spin susceptibility. In *Piloted Aircraft Simulation Techniques*. AGARD-CP-249, Paris, Advisory Group on Aerospace Research and Development.

Gray, T. H. & Fuller, R. R. (1977). Simulator training and platform motion in air-to-surface weapon delivery training. In *Proceedings of the 10th NTEC/Industry Conference, Orlando, Florida, 15–17 November*, pp. 81–90, NAVTRAEQUIPCEN IH-294.

Gum, D. R. (1972). *Modelling of the Human Force and Motion Sensing Mechanisms*. AFHRL-TR-72-54. Dayton, Ohio: Human Resources Laboratory.

Gundry, J. (1976). Man and motion cues. In *Third Flight Simulation Symposium*, London: Royal Aeronautical Society.

Hosman, R. J. A. W. & van der Vaart, J. C. (1980). *Thresholds of Motion Perception and Parameters of Vestibular Models Obtained from Test in a Motion Simulator. Effects of Vestibular and Visual Motion Perception on Task Performance*. Memorandum M-372. Delft: Technische Hogeschool Delft.

Huddleston, H. F. & Rolfe, J. S. (1971). Behavioral factors influencing the use of flight simulators for training. *Applied Ergonomics* **2**(3), pp. 141–8.

Jex, H. R., Jewell, W. R., Magaleno, R. E. & Junker, A. M. (1979). Effects of various lateral-beam washouts on pilot tracking and opinion in the "Lamar" Simulator. In *Proceedings of the Fifth Annual Conference on Manual Control, Wright State University, 20–22 March*. AFFDL-TR-79-3134, pp. 244–66. Dayton, Ohio: Flight Dynamics Laboratory.

Jones, E. R. (1979). The interrelationships between engineering development simulation and flight simulation. In *50 Years of Flight Simulation*, London: Royal Aeronautical Society.

Koevermans, W. P. & Jansen, C. J. (1976). Design and performance of the four-degree-of-freedom motion system of the NLR research flight simulator. In *Flight Simulator/Guidance System Simulation*, AGARD-CP-249, Paris. Advisory Group on Aerospace Research and Development.

Kron, G. J. (1975). *Advanced Simulator for Undergraduate Pilot Training: Motion System Development*. AFHRL-TR-75-59(II). Brooks Air Force Base, TX: Air Force Human Resources Laboratory.

Kron, G. J. & Kleinwaks, J. M. (1978). Development of the advanced G-cuing system. AIAA paper no. 78–1572 in *Flight Simulation Technologies Conference, Arlington TX, 18–20 September, Technical Papers*, pp. 4–14. New York: American Institute of Aeronautics and Astronautics.

Kron, G. J., Cardullo, F. M. & Young, L. R. (1981). *Study and Design of High G Augmentation Devices for Flight Simulation*. AFHRL-TP-80-41. Brooks Air Force Base TX: Human Resources Laboratory.

Lacroix, M. (1979). Recent advances in control loading and motion systems used in simulation. In *50 Years of Flight Simulation*. London: Royal Aeronautical Society.

Lancraft, R., Zacharias, G. & Baron, S. (1981). Pilot/vehicle model analysis of visual and motion cue requirements in flight simulation, AIAA paper no. 81–0972. In *AIAA Flight Simulation Technologies Conference, a Collection of Technical Papers*, pp. 49–59. New York: American Institute of Aeronautics and Astronautics.

Lee, A. H. (1971). Flight Simulator Mathematical Models in Aircraft Design. In AGARD-CP-79-70. Paris, Advisory Group for Aerospace Research and Development.

Levison, W. H. & Junker, A. M. (1978). A model for the pilot's use of motion cues in steady-state roll-axis tracking tasks. AIAA paper no.78–1593 in *Flight Simulation Technologies Conference, a Collection of Technical Papers*. Arlington TX, 18–20 September, pp. 149–59. New York: American Institute of Aeronautics and Astronautics.

McKissick, B. T., Ashworth, B. R. & Parrish, R. V. (1983). An investigation of motion base cueing and G-seat cueing on pilot performance in a simulator. AIAA paper no. 83–1084. In 1983 *Simulation Technologies Conference, Collection of Technical Papers*, pp. 44–51. New York: American Institute of Aeronautics and Astronautics.

Markham, S. R. (1984). *Consideration in the Use of In-Flight and Ground Based Simulations*. AIAA-84-2102-CP. [T110]

Mathews, N. O. & Martin, C. A. (1978). The Development and Evaluation of a g-seat for a high performance military aircraft training simulator. In *Piloted Aircraft Simulation Techniques*. AGARD-CP-249, Paris, Advisory Group on Aerospace Research and Development.

Meiry, J. L. (1966). *The Vestibular System and Human Dynamic Space Orientation*. NASA CR-628. Washington: National Aeronautics and Space Administration.

Mills, G. R. (1967). Improving the quality of motion reproduction in moving-base, piloted flight simulators. *Journal of Aircraft* **4**(5), pp. 439–44.

O'Dierna, A. (1970). *A Synergistic Six Degree of Freedom Motion System*. AIAA paper no. 70–358.

Ormsby, C. C. (1974). *Model of Human Dynamic Orientation*. NASA CR-132537. Washington: National Aeronautics and Space Administration.

Parrish, R. V., Dieudonne, J. E. & Martin, D. J., Jr. (1973). *Motion Software for a Synergistic Six-Degree-of-Freedom Motion Base*. NASA TN D-7350. Washington: National Aeronautics and Space Administration.

Parrish, R. V., Dieudonne, J. E., Bowles, R. L. & Martin, D. J., Jr. (1975). Coordinated adaptive washout for motion simulators. *Journal of Aircraft* **12**(1), pp. 44–50.

Parrish, R. V. & Martin, D. J., Jr. (1976). *Comparison of a Linear and a Nonlinear Washout for Motion Simulators Utilizing Objective and Subjective Data from CTOL Transport Landing Approaches*. NASA TN D-8157. Washington: National Aeronautics and Space Administration.

Parrish, R. V. (1978). Platform motion for fight simulation. AIAA paper 78–1574. In *Flight Simulation Technologies Conference, a Collection of Technical Papers*. Arlington TX, 18–20 September, pp. 21–31. New York: American Institute of Aeronautics and Astronautics.

Perry, D. H. & Naish, J. M. (1964). Flight simulation for research. *Journal of the Royal Aeronautical Society*, **68**(646), pp. 645–62.

Puig, J. A. (1971). The sensory interaction of visual and motion cues. In *Naval Training Devices Center 25th Anniversary Commemorative Technical Journal*, pp. 55–65. Orlando, Florida: Naval Training Equipment Center.

Puig, J. A., Harris, W. T. & Ricard, G. L. (1978). *Motion in Flight Simulation: An Annotated Bibliography*, NAVTRAEQUIPCEN IH-298. Orlando, Florida: Naval Training Equipment Center.

Reynolds, P. A., Schelhorn, A. E. & Wasserman, R. (1973). *Drive Logic for In-Flight Simulators*. AIAA paper no. 73–933.

Reynolds, P. A. (1982). Simulation for predicting flying qualities. In *Criteria for Handling Qualities of Military Aircraft*, AGARD CP-333. Paris, Advisory Group for Aerospace Research and Development.

Ricard, G. L., Parrish, R. V., Ashworth, B. R. & Wells, M. D. (1980). *The Effects of Various Fidelity Factors on Simulated Helicopter Hover*. NAVTRAEQUIPCEN IH-321. Orlando, Florida: Naval Training Equipment Center.

Schmidt, S. F. & Conard, B. (1970). *Motion Drive Signals for Piloted Flight Simulators*. NASA CR-1601. Washington: National Aeronautics and Space Administration.

Sinacori, J. B. (1973). *A Practical Approach to Motion Simulation*. AIAA paper no. 73–931.

Sinacori, J. B., Stapleford, R. L., Jewell, W. F. & Lehman, J. M. (1977). *Researcher's Guide to the NASA Ames Flight Simulator for Advanced Aircraft*, NASA CR 2875. Washington: National Aeronautics and Space Administration.

Sinacori, J. B. (1977). *The Determination of Some Requirements for a Helicopter Flight Research Simulation Facility*. NASA CR 152066. Washington: National Aeronautics and Space Administration.

Staples, K. J. (1978). Current problems of flight simulators for research. *Aeronautical Journal of the Royal Aeronautical Society*, January, pp. 12–32.

Stark, E. A. & Wilson, J. M., Jr. (1973). *Visual and Motion Simulation in Energy Manoeuvring*. AIAA paper no 73–934.

Stark, E. A. (1976). Motion perception and terrain visual cues in air combat simulation. In *AIAA Visual and Motion Simulation Conference, Dayton, Ohio, 26–28 April*, pp. 39–49. New York: American Institute of Aeronautics and Astronautics.

van Gool, M. F. C. (1978). The influence of simulator motion washout filters on pilot tracking performance. In *Piloted Aircraft Simulation Techniques*, AGARD-CP-249, Paris, Advisory Group on Aerospace Research and Development.

Woomer, C. W. & Williams, R. L. (1978). Environmental requirements for simulated helicopter/VTOL operations from small ships and carriers. In *Piloted Aircraft Simulation Techniques*. AGARD-CP-249, Paris, Advisory Group on Aerospace Research and Development.

Young, L. R. (1967). Some Effects of Motion Cues on Manual Tracking *Journal of Spacecraft*, 4(10), 1300–3.

Young, L. R., Meiry, J. L., Newman, J. S. & Feather, J. E. (1969). *Research in Design and Development of a Functional Model of the Human Nonauditory Labyrinths*. AMRL-TR-68-102. Dayton, Ohio: Air Force Aerospace Medical Research Laboratory.

Young, L. R., Oman, C. M., Curry, R. E. & Dichgans, J. M. (1973). *A Descriptive Model of Multi-Sensor Human Spatial Orientation with Applications to Visually Induced Sensations of Motion*. AIAA paper no. 73–915.

Young, L. R., Curry, R. E. & Albery, W. B. (1976). A motion sensing model of the human for simulator planning. In *Readiness Through Simulation, Proceedings of the 9th NTEC/Industry Conference, Orlando, Florida, 9–11 November*. NAVTRAEQUIPCEN IH-276, pp. 149–52. Orlando: Naval Training Equipment Center.

Young, L. R. (1978). Visually induced motion in flight simulation. In *Piloted Aircraft Simulation Techniques*. AGARD-CP-249, Paris, Advisory Group on Aerospace Research and Development.

Zuccaro, J. J. (1970). *The Flight Simulator for Advanced Aircraft – A New Aeronautical Research Tool*. AIAA paper no. 70–359.

Chapter 7

Baldwin, D. M., Goldiez, B. F., Graf, C. P. & Dillinghan, T. W. (1983). *Real-time CGSI – Single Pipeline Processor*. Fifth Interservice/Industry Training Equipment Conference Proceedings, 14–16 Nov, pp. 243–52.

Biberman, L. M., ed. (1973). *Perception of Displayed Information*, pp. 59–65. New York: Plenum Press.

Breglia, D., Spooner, A. & Lobb, D. (1981). *Helmet Mounted Laser Projector.* Image Generation/Display Conference II. Sponsored by US Air Force Human Resources Laboratory, 10–12 June, pp. 241–58.

Breglia, D. (1981). *Helmet Mounted Laser Projector* Interservice/Industry Training Equipment Conference 30 Nov–2 Dec, pp. 8–18.

De Maio, J. (1983). *Visual Cueing Effectiveness; Comparison of Perception and Flying Performance.* Fifth Interservice/Industry Training Equipment Conference, 14–16 Nov, pp. 92–6.

Devarajan, V., Hooks, J. T. & McGuire, D. C. (1984). *Low Altitude High Speed Flight Simulation Using Videodisc Technology.* The 1984 Image Conference III, 30 May–1 June, pp. 53–65. The Air Force Human Resources Laboratory, Williams Air Force Base, Arizona, USA.

Driskell, C. R. & Spooner, A. M. (1976). *Wide-angle Scanned Laser Visual System.* Ninth Naval Training Equipment Center/Industry Conference Proceedings, 9–11 Nov, pp. 97–102.

Farrell, R. J. & Booth, J. M. (1984). *Design Handbook for Imagery Interpretation Equipment* D180-19063–1, Boeing Aerospace Company, Seattle, Washington, USA.

Federal Aviation Administration (1983). *Airplane Simulation and Visual System Evaluation.* Advisory Circular, Federal Aviation Administration, US Department of Transportation.

Gibson, J. J. (1950). *The Perception of the Visual World.* Cambridge, Mass, USA: The Riverside Press.

Gold, T. (1972). *The Limits of Stereopsis for Depth Perception in Dynamic Visual Situation.* Society for Information Display, 1972, International Symposium, Digest of Technical Papers, pp. 150–51.

LaRussa, J. A. (1964). *The Infinity Image System in Visual Simulation.* American Institute for Aeronautics and Astronautics, November, pp. 263–70.

Schacter, B. J. (1983). *Computer Image Generation.* USA: John Wiley.

Spooner, A. M. (1976). Collimated displays for flight simulation. *Optical Engineering,* May–June, pp. 215–19.

Spooner, A. M. (1979). Visual simulation – past, present and future. In *Fifty Years of Flight Simulation Conference,* April, London, Royal Aeronautical Society.

Spooner, A. M., Breglia, D. R. & Patz, B. W. (1980). *Realscan – A CIG System with Greatly Increased Image Detail,* Second Interservice/Industry Training Equipment Conference.

Spooner, A. M. (1982). *The Trend Towards Area of Interest In Visual Simulation Technology.* Interservice/Industry Training Equipment Conference, 16–18 Nov, pp. 205–15.

Sutherland, I. E., Sproule, R. F. & Schumacker, R. A. (1974). *A Characterisation of Ten Hidden Surface Algorithms. ACM Computing Surveys,* Vol. 6, No. 1, March, pp. 1–55.

Szabo, N. S. (1978). *Digital Image Anomalies; Static and Dynamic.* Proceedings of the Society of Photo-Optical Instrumentation Engineers, Visual Simulation and Image Realism, Vol. 162, pp. 11–15.

Tong, H. M. (1983). *A Laser Image Generation System for Helicopter Nap-of-the-Earth Flight Training.* Fifth Interservice/Industry Training Equipment Conference.

Tong, H. M. & Fisher, R. A. (1984). *Progress Report on an Eye-slaved Area-of-interest Visual Display.* The 1984 Image Conference III, 30 May–1 June, The Air Force Human Resources Laboratory, Williams Air Force Base, Arizona, USA, pp. 279–94.

Webb, P. ed. (1964). *Bioastronautics Data Book.* NASA Sp-3006. NASA: Washington. D.C., USA.

Chapter 8

Adams. J. A. (1979). On evaluation of training devices. *Human Factors*, **21** (6), pp. 711–20.

Barnes, A. G. (1984). Application of air combat simulation to pilot training. In *Future Applications and Prospects for Flight Simulation*, London: Royal Aeronautical Society.

Bolton, M., Campbell, D., Murray, P., Olive, G. & Roberts, M. (1979). Recent and future engineering developments in flight training simulators. In 50 *Years of Flight Simulation*, London: Royal Aeronautical Society.

Dickman, J. L. (1984). Training management systems: an assessment of current status and future potential. Spring Convention, *Future Applications and Prospects for Flight Simulation*, May. London: Royal Aeronautical Society.

Ebeling, F. A., Goldhor, R. S. & Johnson, R. L. (1972). A scanned infra red light beam touch entry system. *Society for Information Display International Symposium*, pp. 134–5 Los Angeles: Society for Information Display.

Gorrell, E. L. (1980). *A human engineering specification for legibility of alphanumeric symbology on video monitor displays.* DCIEM Tech. Report No. 80-R-26, Department of National Defence, Canada.

Mackey, C. (1980). *Human factors aspects of visual display unit operation.* Health and Safety Executive, London: HMSO.

MIL STD 1472C (1981). *Military Standard. Human engineering design criteria for military systems, equipment and facilities.* US Government Printing Office, 2 May.

Morrell, A. M. (1981). Color picture tube design trends. *Proceedings of the Society for Information Display*, Vol. 22 No. 1, pp. 3–9. Los Angeles: Society for Information Display.

Pearson, D. E. (1975). *Transmission and display of pictorial information.* Pentech Press Ltd, New York: John Wiley.

Rupp, B. A. (1981). Visual display standards: a review of issues. *Proceedings of the Society for Information Display.* Vol. 22 No. 1, pp. 63–72. Los Angeles: Society for Information Display.

Spencer, G. R. (1981). Performance of penetration color crts in single-anode and dual-anode configurations. *Proceedings of the Society for Information Display.* Vol. 22 No. 1. pp. 15–17. Los Angeles: Society for Information Display.

Spring, W. G. (1976). Advanced flight simulation in air combat training. American Institute of Aeronautics and Astronautics Conference Proceedings, *Visual and Motion Simulation*, Dayton, Ohio, April.

Umbers, I. G. (1977). *A review of human factors data on input devises used for process computer communication.* Warren Spring Laboratory, DOI, Hertfordshire, UK.

Van Cott, H. P. & Kinkade, R. G. (1972). *Human engineering guide to equipment design.* US Government Printing Office.

Chapter 9

Adams, J. A. (1979). Evaluation of Training Devices. In *50 Years of Flight Simulation*, London, Royal Aeronautical Society.

CCA (1984). Draft CAP 453. London, UK Civil Aviation Authority.

Davies, D. P. (1968). *Handling the Big Jets*, Brabazon House, Redhill, Surrey, UK, Air Registration Board.

Davies, D. P. (1975). *Approval of Flight Simulator Flying Qualities.* London, Aeronautical Journal of the Royal Aeronautical Society, July.

DOA (1982). *Advisory Circular No. FWD* 1/1982, Sydney, Australia, Department of Aviation.

FAA (1980). *Federal Aviation Requirement* 121 *Appendix H.* Washington, USA, Federal Aviation Administration.

FAA (1983). *Federal Aviation Advisory Circular AC* 120–40. FAA Southern Region,

Atlanta, Federal Aviation Administration.

Ferrarese, J. P. (1979). Criteria for Approval of Flight Training Simulators. In *50 Years of Flight Simulation*, London, Royal Aeronautical Society.

MIL STD 1644A (1982). *Military Standard Trainer System Software Engineering Requirements*. Washington, USA, Government Printing Office.

Transport Canada (1982). *Manual of Aircraft Simulation Approval*. Quebec, Canada, Canadian Government Publishing Centre.

Chapter 10

AGARD (1978). *Piloted Aircraft Environment Simulation Techniques*, AGARD-CP-249, Paris, Advisory Group on Aerospace Research and Development.

AGARD (1980). *Fidelity of Simulation for Pilot Training*, AGARD-AR-159, Paris, Advisory Group on Aerospace Research and Development.

Barnes, A. G. (1967). *A Simulator Investigation of Rolling Requirements for Landing Approach*, ARC R and M 3605.

Cooper, G. E. & Harper, R. P. Jr. (1969). *The Use of Pilot Rating in the Evaluation of Aircraft Handling Qualities*, NASA TN D-5153.

Jones, E. (1979). In *Fifty Years of Flight Simulation*, London: Royal Aeronautical Society.

Neal, P. T. & Smith, R. E. (1970). *An In-Flight Investigation to Develop Control System Design Criteria for Fighter Airplanes*, AFFDL-TR-70-74.

White, R. G. & Beckett, P. (1983). *Increased Aircraft Survivability Using Direct Voice Input*. In AGARD-CP-347, Paris, Advisory Group on Aerospace Research and Development.

Zaitzeff, L. P. (1969). Aircrew task loading in the Boeing Multi-Mission Simulator. In *The Measurement of Aircrew Performance*, AGARD-CP-56, Paris, Advisory Group on Aerospace Research and Development.

Chapter 11

AGARD (1980). *Fidelity of Simulation for Pilor Training*. AGARD Report No. 159, Paris, Advisory Group on Aerospace Research and Development.

Barnes, A. G. (1984). Applications of air combat simulation to pilot training. In *Future Applications & Prospects for Flight Simulation*. London: Royal Aeronautical Society.

Bird, J. D. (1978). Design concepts of the Shuttle Mission Simulator. In *Extending the Scope of Flight Simulation*. London: Royal Aeronautical Society.

Blaiwes, A. S., Puig, J. A. & Regan, J. J. (1973). Transfer of training and the measurement of training effectiveness. *Human Factors*, **15**, 523–33.

Cream, B. W., Eggemeier, F. T. & Klein, G. A. (1978). A strategy for the development of training devices. *Human Factors*, **20**, 145–58.

Dickman, J. L. (1983). A comparative analysis of two basic philosophies in instructors' station design. In *Simulators*. London: Institute of Electrical Engineers.

Dickman, J. L. (1984). Training management systems: an assessment of current status and future potential. In *Future Applications & Prospects for Flight Simulation*. London: Royal Aeronautical Society.

Durose, C. G. (1982). An evaluation of some experimental data on the cost effectiveness of flight simulators. *The Aeronautical Journal*, March, pp. 90–3.

Eddowes, E. E. & Waag, W. L. (1980). *The Use of Simulators for Training In-Flight Emergency Procedures*. AGARDograph No. 248. Paris, Advisory Group on Aerospace Research and Development.

Ferrarese, J. A. (1978). Criteria for the approval of flight training simulators. In *Fifty Years of Flight Simulation*. London: Royal Aeronautical Society.

Gillman, R. E. (1969). In defence of flight simulators. *Flight International*, **96**, 93–4.

Hammerton, M. (1977). Transfer and simulation. In *Human Operators & Simulation*. London: Institute of Measurement & Control.

Holman, G. L. (1979). *Training Effectiveness of the CH-47 Flight Simulator*. US Army Research Institute for the Behavioural Sciences Report No 1209. Alexandria Va.

Houston, R. C. (1984). Phase II is enough. In *Future Applications & Prospects for Flight Simulation*. London: Royal Aeronautical Society.

Huettner, C. E. (1984). The advancement of advanced simulation. In *Future Applications and Prospects for Flight Simulation*. London: Royal Aeronautical Society.

Hughes, R. L., Brooks, R., Graham, D., Sheen, R. & Dickens, T. (1983). Tactical ground attack: on the transfer of training from flight simulator to operational Red Flag range exercise. In *Simulators*. London: Institute of Electrical Engineers.

Hussar, J. J. (1983). A comparison of simulator procurement/program practices: military vs. commercial. In *5th Interservices/Industry Training Equipment Conference*. Washington D.C.

Johnson, W. J. (1968). Flight simulation and airline training. In *Vehicle Simulation for Training & Research*. RAF Institute of Aviation Medicine Report No. 442.

Jones, E. R. (1978). The interrelationships between engineering development simulation and flight simulation. In *Fifty Years of Flight Simulation*. London: Royal Aeronautical Society.

Kelly, L. L. & Parke, B. P. (1970). *The Pilot Maker*. New York: Grosset & Dunlap.

Knight, M. A. G. (1982). Low cost aircrew training devices. *The Aeronautical Journal*. March, pp. 98–101.

Martz, J. W. (1984). The changing role of supplemental simulators. In *Future Applications & Prospects for Flight Simulation*. London: Royal Aeronautical Society.

Miller, R. M., Swink, J. R. & McKenzie, J. F. (1979). *Instructional Systems Development in Air Force Flying Training*. Report No AFHRL-TR-78-59. Williams Air Force Base Arizona. Flying Training Division, Air Force Human Resources Laboratory.

Orlansky, J. (1985). *The Cost-Effectiveness of Military Training*. Alexandria, Virginia: Institute for Defense Analyses.

Orlansky, J. & Chatelier, P. R. (1983). The effectiveness and cost of simulators for training. In *Simulators*. London: Institute of Electrical Engineers, Conference Publication No 226.

Osgood, C. E. (1949). The similarity paradox in human learning: a resolution. Psychology Review, **56**, pp. 132–43.

Povenmire, H. K. & Roscoe, S. N. (1973). Incremental transfer effectiveness of a ground based general aviation trainer. *Human Factors*, **15**, 534–42.

Rolfe, J. M. & Caro, P. W. (1982). Determining the training effectiveness of flight simulators: some basic issues and practical developments. *Applied Ergonomics*, **13**, 243–50.

Rolfe, J. M. & Riglesford, W. B. (1984). The application of cardboard technology to flying training. *RAF Education Bulletin* No. 22.

Roscoe, S. N. (1971). Incremental transfer effectiveness. *Human Factors*, **13**, 561–7.

Roscoe, S. N. (1980). *Aviation Psychology*. Ames Iowa: Iowa State University Press.

Royal Aeronautical Society (1977). Flight simulator instructor training. *The Aeronautical Journal*, December, pp. 15–527.

University of Illinois (1979). *A Decade with the Institute of Aviation*. Urbana-Champaign: University of Illinois.

Wooden, W. A. (1978). An assessment of the state of training simulators. In *Fifty Years of Flight Simulation*. London: Royal Aeronautical Society.

Chapter 12

Ashworth, B. R., McKissick, B. T. & Parrish, R. V. (1984). *Effects of Motion Base and G-Seat on Simulator Pilot Performance*, NASA TP 2247.

Borah, J., Young, L. R. & Curry, R. E. (1979) *Sensory Mechanism Modelling* AFHRL-TR-78-83. Dayton Ohio: Human Resources Laboratory.

Federal Aviation Administration. (1983). *Airplane Simulation and Visual System Evaluation.* Advisory Circular, Federal Aviation Administration, US Department of Transportation.

Fraser, T. M. (1966). *Philosophy of Simulation in a Man–Machine Space Mission System,* NASA SP 102.

Morgan, R. L. (1971). *Implications of Training for CAI,* AFHRL-TR-71-35.

Nashner, L. M. (1970). *Sensory Feedback in Human Posture Control,* Massachusetts Institute of Technology, MVT-70-3.

Palmer, E. P. & Petitt, J. (1976). Differences for judgement of sink rate during flare. In *AIAA Visual and Motion Simulation Conference,* April.

Parrish, R. V. Dieudonne, J. E., Bowles, R. L. & Martin, D. J., Jr. (1975). Coordinated adaptive washout for motion simulators. *Journal of Aircraft* **12,** 44–50.

Parrish, R. V. & Martin, D. J. Jr. (1976). *Comparison of a Linear and a Nonlinear Washout for Motion Simulation Utilizing Objective and Subjective Data from CTOL Transport Landing Approaches* NASA TN D-8157. Washington: National Aeronautics and Space Administration.

Parrish, R. V. & Steinmetz, G. C. (1983). *Evaluation of G-Seat Augmentation of Fixed Base/Moving Base Simulation for Transport Landings Under Two Visually Imposed Runway Width Conditions,* NASA TP 2135.

Showalter, T. W. & Parris, B. L. (1980). *The Effects of Motion and G-Seat Cues on Pilot Simulation Performance of Three Piloting Tasks,* NASA TP 1601.

Thanderz, M. (1982). Assessing pilot performance and mental workload in training simulators. In *Simulation – Avionic Systems and Aeromedical Aspects,* London: Royal Aeronautical Society.

Index